PENGUIN CLASSICS

THE NATURE OF THINGS

TITUS LUCRETIUS CARUS must have been born soon after 100 BC and is likely to have died before his poem was given to the world, probably in the 50s BC. Almost nothing is known about his life. He was a Roman citizen and a friend of Gaius Memmius, a Roman politician, and his poem was read and admired by Cicero. It is doubtful if there is any truth in the story preserved by St Jerome and immortalized by Tennyson that he died at his own hand after being driven mad by a love philtre.

A. E. STALLINGS was born in 1968. She studied Latin at the University of Georgia, and Latin and Greek at Oxford University. She has published two collections of poetry, *Archaic Smile*, which won the Richard Wilbur Award, and *Hapax*. Her work has twice been included in the Best American Poetry series, and has received a Pushcart Prize among other awards. She lives in Athens, Greece, with her husband, the journalist John Psaropoulos, and their son, Jason.

RICHARD JENKYNS was born in 1949 and educated at Eton and Balliol College, Oxford. He was a Fellow of All Souls College, Oxford, from 1972 to 1981, and a lecturer at the University of Bristol between 1978 and 1981. He has been a Fellow of Lady Margaret Hall, Oxford, since 1981, and Professor of the Classical Tradition since 1999.

LUCRETIUS

The Nature of Things

Translated and with Notes by
A. E. STALLINGS
Introduction by RICHARD JENKYNS

PENGUIN BOOKS

PENGUIN CLASSICS

Published by the Penguin Group
Penguin Books Ltd, 80 Strand, London WC2R 0RL, England
Penguin Group (USA) Inc., 375 Hudson Street, New York, New York 10014, USA
Penguin Group (Canada), 90 Eglinton Avenue East, Suite 700, Toronto, Ontario, Canada M4P 2Y3
(a division of Pearson Penguin Canada Inc.)
Penguin Ireland, 25 St Stephen's Green, Dublin 2, Ireland
(a division of Penguin Books Ltd)
Penguin Group (Australia), 250 Camberwell Road, Camberwell, Victoria 3124, Australia
(a division of Pearson Australia Group Pty Ltd)
Penguin Books India Pvt Ltd, 11 Community Centre, Panchsheel Park, New Delhi – 110 017, India
Penguin Group (NZ), 67 Apollo Drive, Rosedale, North Shore 0632, New Zealand
(a division of Pearson New Zealand Ltd)
Penguin Books (South Africa) (Pty) Ltd, 24 Sturdee Avenue, Rosebank, Johannesburg 2196, South Africa

Penguin Books Ltd, Registered Offices: 80 Strand, London WC2R 0RL, England

www.penguin.com

Published in Penguin Classics 2007

011

Translation and Notes copyright © A. E. Stallings, 2007
Introduction and Further Reading copyright © Richard Jenkyns, 2007
All rights reserved

The moral rights of the translator and editor have been asserted

Set in 10.25/12.25 pt PostScript Adobe Sabon
Typeset by Rowland Phototypesetting Ltd, Bury St Edmunds, Suffolk
Printed in England by Clays Ltd, St Ives plc

ISBN: 978-0-140-44796-5

www.greenpenguin.co.uk

ALWAYS LEARNING   **PEARSON**

Contents

Introduction

Of all the great poems of Europe – and it is indeed among the greatest – Lucretius' *De Rerum Natura* (*The Nature of Things*) is perhaps the most improbable. Here is a poem without people in it, without any story; instead, it offers a treatise on science and philosophy. The philosophy, moreover, is a strict materialism, which denies the existence of anything magical, mysterious or transcendent. It does not sound like promising matter for poetry at all, let alone for a work of more than 7,000 lines. Yet the result is a masterpiece. A key to appreciating this most unlikely success is to understand the nature of Lucretius' beliefs and the circumstances in which he decided to expound them.

About Titus Lucretius Carus himself we know almost nothing. A very few supposed facts about his life come down to us from antiquity, but they appear in sources many centuries later than his time, and they seem to be at best guesswork, at worst pure fabrication. Only one firm date can be attached to him: in 54 BC Cicero slipped a sentence into a letter to his brother praising 'the poems of Lucretius' – presumably the poem that we have or some version of it – for their combination of imagination and skill. Some verbal echoes indicate that Catullus was reading Lucretius while he was writing his own longest poem – Poem 64, commonly known as *Peleus and Thetis* – datable to somewhere in the fifties BC. Though *The Nature of Things* is usually supposed to have been written in the fifties, it has been dated to the sixties, and even to the forties. While it is possible that Lucretius was still tinkering with it at the start of the forties, the bulk of it must be earlier. Since there is no trace of him having written anything else, it may have been a lifetime's work, composed over a decade or more, with portions of it circulating while it was still incomplete. The poem is indeed unfinished. For example, a passage from the first book reappears at the beginning of the fourth, evidence that Lucretius had not finally decided where to place it. And the last book breaks off abruptly in the middle of an account of the plague which swept through Athens in 430 BC.

The fact which we can most confidently assert about Lucretius is one drawn from his poem itself: that he had been converted to the philosophy of Epicurus (341–270 BC) and was possessed with the desire to persuade others of its truth. To understand Lucretius we need to consider what Epicurus taught, and to understand Epicurus we need to set him in his context within the development of Greek thought.

One of the achievements of the Greek mind between the eighth and the fifth centuries was a process which might be called separation or differentiation. They discovered that fact is different from fiction and that history is different from myth, that theology and philosophy are different ways of talking about the world, and that each of these is different from natural science. These distinctions may seem very obvious to us, but they were not then. Take Homer's *Iliad* and Hesiod's *Theogony*, poems datable probably to the later eighth century. Homer was both poet and historian, and Hesiod saw no firm distinction between story-telling about the origin of the gods and explaining how the world came into being in physical terms. Similarly, we may find it hard to decide whether the Greek intellectuals of the seventh and sixth centuries are best described (in our terms) as philosophical or as scientific thinkers. But gradually the discovery that science and philosophy are different modes of enquiry was made, and in the later fifth century Socrates made the distinction absolute.

This distinction has prevailed in our thought ever since. To put the matter in very simplified terms: the dominant tradition of Western philosophy descending from Socrates' pupil Plato and Plato's pupil Aristotle tells us to look inwards and work out the nature of reality within our own minds by a process of abstract reasoning. By contrast, Epicurus held that natural science is the route to philosophical understanding, and his system, from one point of view, can be seen as the revival, in a transformed shape, of an older tradition of Greek thought. He begins not with abstract cogitation but with the external world. On his account we can deduce from the physical phenomena around us that all matter is made up of small indivisible particles – atoms – and that nothing exists except atoms and empty space. No atom can be created and none can be destroyed; space is infinite, as is the number of atoms that move within it.

All events and processes are merely the effects of the movement of atoms. All existence is material; everything that exists is part of nature, and therefore there can be no supernatural realm. It is disputed in what sense Epicurus believed in the existence of the gods: the usual view is that he believed the gods to exist as independent entities, but some think that he believed them to exist merely as concepts. This much is sure: insofar as the gods exist, they must be made of atoms, like everything else; they did not create the world, they play no part in its governance and they take no interest in us.

Epicurus argues that certain moral truths follow necessarily from these scientific facts. No one can rationally pursue anything other than his own pleasure. Epicurus is thus in the strict sense a hedonist. However, he places a fairly low value on pleasure as the *homme moyen sensuel* or man in the street is likely to conceive it. Hunger, thirst and sexual desire are necessary appetites, but we should try to moderate them as far as our natures allow. Romantic love, for example, is to be avoided, as it involves a loss of rationality and self-control (Lucretius makes this argument in the later part of his fourth book). All pleasures of the senses are inferior to such abstract pleasures as friendship and philosophical contemplation. Accordingly, this philosophy, often caricatured as a religion of sensuous self-indulgence, is in reality rather austere. No one could be less of an epicure than Epicurus.

He also offers to free us from anxiety about death. This is the argument worked through in Lucretius' third book. Mind and soul are purely material, and death is merely the dissolution of a temporary combination of atoms. 'Then Death is nothing to us', says the poet, translating the first of his master's *Sovereign Maxims* (III.322). Liberated from this fear, and converted to the truths of

Epicureanism, we can live in serene and secure pleasure – as Lucretius says in another place, we can have a life worthy of the gods.

The philosophies of the Hellenistic age (that is, the period of three centuries following the death of Alexander the Great in 324 BC), such as Epicureanism and its rival Stoicism, founded by Zeno of Citium (335–263 BC), offer mankind salvation. In other terms, they perform some of the functions that in our world are performed by religious belief. We can divide people into those who believe in the existence of a god or gods, and those who do not; and today we usually expect believers to have a more hopeful view of life than unbelievers. There is another division, however, which may seem similar but is in fact different; this is between those who believe that the order of things offers some ultimate comfort or safety, and those who have no such confidence. The pious Sophocles (c. 495–406) believes in the gods, but seems to recognize that there is no horror so great that they will protect us from it. The materialist Epicurus, by contrast, denies the existence of providence, but nevertheless holds that we can be assuredly safe. Both Stoicism (which does believe in a divine providence) and Epicureanism assert that the mind of the wise man, enlightened by philosophy, is an impregnable citadel. This is the doctrine that Lucretius expounds at the start of his second book. It is pleasant, he says, to watch a boat struggling at sea while one is safe on shore, or to look down from one's philosophical height upon armies clashing in the plain below – not because the wise man is a sadist, but because his mind is impassible, and no torment can touch it. The goal of Stoicism was *apatheia*, not feeling, the goal of Epicureanism *ataraxia*, not being disturbed – both negative ideals. If we were to look for an analogy in modern spirituality, we might turn to Buddhism sooner than to Christianity.

Lucretius distributes three grand passages in praise of Epicurus symmetrically across his poem. After the hymn to Venus that opens the first book, he celebrates his master as a man; at the start of the third book he praises him as a father, and at the start of the fifth as a god. Epicurus himself might not have disapproved: he seems to have wanted his followers to form a kind of 'church', encouraging them to share their lives together. He even appears to have devised a kind of Sunday and a kind of Christmas: once a month he enjoined his disciples to gather together, and once a year, on his birthday, to commemorate their founder. When Lucretius declares, in the fifth book, 'He was a god, a god indeed' (V.8), we can take that as a metaphor, even though the poet is describing the saving power of his master's teaching. More remarkable, perhaps, is Lucretius' description, in the third book, of the divine delight and shiver that grip him when he contemplates the master's words (III.28–9), for that combination of joy and something like fear seems to evoke a religious sense of the numinous.

There is an obvious paradox in calling Lucretius religious. His denial of the supernatural is absolute, he extols Epicurus for trampling religion underfoot, he declares that religion has often led people into wicked and cruel acts and he illustrates this with the story of Iphigenia, killed by her father Agamemnon as a sacrificial victim so that the Greeks might obtain a fair wind for their voyage to Troy. He sums this up in a line as famous as any in the entire poem: 'So potent was Religion in persuading to do wrong' (I.101). And yet it is hardly less obvious that his work is suffused with religious language, and he opens it with a hymn to Venus – the most magnificent and spectacular act of worship in classical Latin literature. What is his purpose?

Scholars have commonly been puzzled or embarrassed by this opening. Some even suggest that Lucretius means to mislead his readers, luring them into the poem by a false idea of the gods which will later be corrected. But it should be plain that this glorious beginning sets the tone of the poem as a whole, that it expresses in metaphorical form his most intensely held beliefs, and that it is central to his purpose that he invites the reader, from the very first line, to a posture of adoration. Earlier poets had found the gods within or behind or beyond nature; Lucretius uses and yet transforms this tradition by coming in a spirit of worship to the facts of nature themselves. It is mere atoms and the movement of atoms that we are to adore.

Lucretius' Venus may seem to have too many functions: she is the mother of the Romans, the Epicurean pleasure principle, the season of spring, the sexual drive, the goddess of peace and a kind of muse invoked to impart beauty to the poet's language. How, it may be asked, can a single figure symbolize so many disparate things? But that is precisely Lucretius' point: everything that happens – the experience of pleasure, the winds blowing, the flowers blooming in the spring, the beasts rutting, the poet composing – has the same essential cause. Every action, all creation and all destruction are alike the product of the push and pull of atoms, of these elementary particles colliding, cohering or flying apart.

Lucretius brings his hymn to its climax with a picture of the sexual union between Venus and Mars, the god of war. This again may seem baffling. Some of the poet's language suggests harmony, balance and equality between the two deities; some of it suggests peace prevailing over war, Mars yielding and Venus victorious. But this paradox too contains a profound meaning. Lucretius is drawing on a symbolism used by Empedocles, a fifth-century cosmologist-cum-philosopher who wrote in verse and whose poetry Lucretius praises later in the first book. Empedocles held that the universe was governed by an interplay between two forces, Love and Strife, and he gave his theory a mythological dress by calling these forces Ares and Aphrodite. These two deities became lovers in a famous story told in the eighth book of Homer's *Odyssey*; their Roman equivalents are Mars and Venus. Strife or Ares, it appears, is the force that dissevers, Aphrodite or Love the force that unites. Both, it would seem, are necessary for the world to flourish: without Love there would be no coherence or continuity, without Strife no activity or new creation. So Love and Strife need to be in balance or harmony. But the universe as totality needs to cohere, and so the sum of things, we may infer, is the prevailing of Love over Strife.

Such, at least, seems to be Lucretius' idea. Epicurus himself had the reputation of being a dry and crabbed writer, with no care for the poetry of things, but his Roman disciple finds the romance in his conception. On the largest and the smallest scale there is immutability: the universe is the sum of infinite space and an infinite number of atoms, and nothing can be added to it or subtracted from it; and each single atom is immortal, indivisible and indestructible. In between, everything is activity and mutability, for the atoms are in ceaseless energetic motion. The drama lies in this combination of change and changelessness, and

Lucretius' erotic metaphor celebrates this grand paradox. The sexual act can be described both as union and as an experience of conquest and surrender; Venus both unites with Mars and conquers him; the universe is on one view a balance of strife and love, peace and war, while on another it is the sum of innumerable lesser strifes within an ultimate and unassailable peace or love.

There is power and splendour in this vision of the world, but one may still wonder how far Lucretius' master would have approved it. Epicurus seems to have had a rather low opinion of poetry, and it is natural to ask why his faithful follower should be so keen to poeticize his teachings. Lucretius gives one reason openly: as we smear honey round the rim of the cup to disguise a medicine's bitter taste, so the charm of verse can make difficult or rebarbative material seem less unappealing. But this suggests that the poem's literary quality is distinct from most of its content, whereas most readers feel that the poetry and the philosophy in Lucretius form an indivisible whole. Part of his purpose, indeed, may have been to persuade us that Epicureanism is, after all, a poetical creed. It was easy to feel that as a materialist philosophy Epicureanism sucked the glory, wonder and mystery out of the world, and that as a philosophy of pleasure it was ignoble: Cicero sometimes talks as though whatever the intellectual case its adherents might make, it was no philosophy for a Roman gentleman. Lucretius' argument in itself aims to show us that the teachings of Epicurus are true; the manner in which he puts that argument aims to show that they are also beautiful and lovable.

There is a deeper reason, however, why Lucretius needs to do his philosophizing in verse: beneath the overt argument another implicit argument runs as an undercurrent, and this implicit argument attempts to meet two of the greatest difficulties of the Epicurean system. These are the problem of death and the problem of altruism. The Epicurean claim about death is a very strong one: it goes beyond saying that death should not be feared, declaring it to be a matter of pure indifference. This runs counter to ordinary intuition: if life is good, and death is the end of life, death would seem to be an evil, even though it may be an evil that we should face calmly. Epicurus, however, asserts that it is not an evil at all.

The problem of altruism is that Epicureanism gives each one of us no reason to consult anything other than our own pleasure; and yet Lucretius' project seems plainly altruistic: he wants to convert and save us. And in this he was, once again, faithful to his master: Epicurus' purpose was not only to achieve happiness for himself but to lead others to the same goal.

It may be that these problems are insoluble, but Lucretius' poetry can be seen as a powerful and ambitious attempt to overcome them while remaining true to his master's teaching. In the poem's very first line Venus is presented as a mother, and in the first passage of philosophical argument Lucretius displays as part of the beauty and diversity of the world the fact that everything has a fixed mother. A little later, in another example of religious imagery, he borrows from Greek fertility cult the idea of the 'sacred marriage', with Father Sky sending down the life-giving rain to impregnate Mother Earth (I.250–51). Thanks to this vivifying act, he continues, the crops and trees grow, the birds and beasts give birth to their young, and – a more surprising twist – 'we see happy cities all abloom with girls and boys' (255). Nature and culture, plants, animals and cities, all alike are the products of one energizing force; or in the terms of Epicurean physics, the activity of atoms is the cause of everything that exists.

Lucretius returns to this imagery of fatherhood and motherhood late in his second book, in an extraordinary passage that to our ears may seem to have an almost scriptural ring:

> Finally, we all arise from seed celestial,
> Because the same sky overhead is father of us all . . .
> And that which was sent down to earth from heaven's
>   aethereal shore
> Is taken up again into the quarters of the sky.
>
> (II.991–1001)

Lucretius then returns to the idea of the pregnant earth giving birth to all life, asserting that therefore the earth is rightly called mother. He will come back to this assertion, repeatedly and with much emphasis, in the fifth book:

Therefore, again and again I tell you, that when men acclaim
The earth as 'Mother Earth', she is deserving of the name . . .

(V.821–2)

These metaphors of parenthood present everything that is, including ourselves, as part of a universal kinship. They contribute to a way of apprehending the world which metaphorically animates every part of it through a nexus of imagery. For example, Epicurus' word for atom was *atomos*, a purely descriptive term, meaning 'that which cannot be divided', but Lucretius calls these particles 'bodies', 'seeds' or 'generative seeds', vivifying and even sexualizing the basic elements of matter. A famous passage early in the second book adds to these ideas two more: the idea of mutual cooperation between all that exists and the idea of eternal newness. The ageing and dying of things, he explains, is the creation of other things, and 'Thus the Sum of Things is every hour / Renewed' (II.74–5). In the old mythology, immortal youth and loveliness were the privilege of the gods; now, in Lucretius' vision, our own world enjoys this blessing. He adds that all things exist in a relationship of mutual give and take, and compares generation succeeding generation to relay runners in a race passing on the torch of life. The significance of this simile is that the runner in a race wants to hand on the torch to his team-mate. The poet hints at an idea that perhaps he cannot quite speak directly: it is as though we too are members of a team, and might come to see our own extinction as fitting, because it is a necessary condition for the everlasting youth and freshness of the world – a world to which we are bound by natural ties of affection as members of a universal family, sharing one father and mother.

What Lucretius has in effect done is to bring to the doctrines of Epicurus a disposition of mind and feeling more characteristic of the rival school of Stoicism. Thus the Stoic Marcus Aurelius (AD 121–80), writing a couple of hundred years later than Lucretius, declares, 'By the changes of the parts of universal nature the whole world continues ever young . . . Now what tends to the advantage of the Whole is ever altogether lovely and in season; therefore for each individual the cessation of his life is no evil.' If we feel something hard to bear, Marcus continues, we have forgotten 'the great kinship of man with all mankind' (*Meditations* XII. 23, 26). In bringing this style of feeling to Epicureanism, Lucretius' originality is both imaginative and intellectual. Indeed, his originality lies in his demonstration that poetic imagination can do more than decorate his philosophy: it can knit its fibres into the philosophy's very substance. Epicurus tells us that we can rationally pursue only our own pleasure. But the affection of parents and children for one another is a natural and instinctive pleasure, rooted in our animal being. Once we come to love the world as the members of a family love one another, we can explain, in Epicurean terms, how it can be rational for us to care for creatures other than ourselves; and we have new cause to accept, and even in a way to desire, our own deaths.

Here is the deepest reason why Lucretius' philosophy must be poetry also. You cannot command a person into a feeling: it may be the best possible thing for Jack to fall in love with Jill, but we cannot instruct him to do so. Similarly, Lucretius cannot instruct us to love nature, the world and the everlasting atomic flux, to be rapt by the romance of a universal kinship; he can only achieve his end by writing a masterpiece so powerful in its poetry that we are persuaded to feel the romance and drama of his conception for ourselves. And therefore he embarked on the eccentric scheme of doing philosophy – real philosophy – in verse.

For it was indeed a highly eccentric project, the loneliest of Latin poems. To see why, we must once again set him in a broad context. In early Greece, verse could be a medium for exposition and argument, as well as for storytelling or lyrics about love; the Athenian statesman Solon (died *c.* 560 BC), for example, promoted his cause in elegiac poems. Some of the Presocratic thinkers expounded their theories in prose, but others, such as Parmenides (born *c.* 515 BC) and Lucretius' own favourite Empedocles (*c.* 492–432 BC), used hexameter verse. However, as part of that process of separation or differentiation which marks the development of their thought, the Greeks came to see that prose and verse had distinct functions; and by the end of the fifth century prose had become the only medium for discursive or intellectual subjects such as history or philosophy. From this time on, and for the rest of European history, almost all didactic poems will be pseudo-didactic; that is to say, although they go through the motions of providing information or instruction, their true purpose is not informing or instructing but giving literary pleasure. Consider the didactic poems of the Hellenistic age: if you were bitten by a snake, you did not consult the *Alexipharmaca* of Nicander (*fl. c.* 130 BC), you called for a doctor; if you really wanted to learn about astronomy, you read a prose treatise, not the *Phaenomena* of Aratus (born *c.* 315 BC). No one supposes that Virgil genuinely meant his *Georgics*, a poem inspired by Lucretius' example, to instruct countrymen how to grow crops and vines, breed stock and keep bees, and any farmer who tried to use it as a handbook would quickly find it inaccurate and incomplete. But Lucretius is genuinely trying to write philosophy in verse, to expound, argue and persuade, and no one, it seems, had seriously attempted that for nearly four hundred years.

The oddity of Lucretius is confirmed by the extent to which other Roman philosophers ignored him. Cicero, despite admiring him as a poet, never once refers to him in his many philosophical works, and does not even pay him the compliment of being worth refutation. Among the writers of antiquity there is perhaps only one who appreciates him as a thinker, and that is the man who had pondered him most deeply: Virgil, offering an eloquent tribute in the second book of the *Georgics*, praises him for the force of his intellect and the strength of his understanding. It is significant that a great poet is the one to value the philosophy in Lucretius: pondering the nature of this masterpiece, Virgil came to recognize, it would seem, that Lucretius' moral seriousness and intellectual energy are the essence of his effect.

Lucretius' poem can be seen as a fusion of two genres, and this very fusion also becomes part of his meaning. The Greeks seem to have applied the term *epos* to any poem composed in hexameter verse. Nevertheless, in practice the Greeks and Romans distinguished two streams of influence, one descending from Homer's *Iliad* and *Odyssey* – what we call epic – and the other descending from Hesiod's *Works and Days* – the genre that we call didactic. Its content makes *The Nature of Things* a didactic poem: quite simply, it is a poem that teaches. But in many ways its form and manner are more like epic. There is the sheer scale of the work, for one thing: about three-quarters the length of Virgil's *Aeneid*, it is far longer than any other didactic poem of antiquity. It is significant that Ennius (239–169), the first great epicist in Latin, and Homer are the two poetic names that Lucretius invokes in his introduction; Homer will recur near the end of the third book, as Lucretius invites us to accept our mortality by reflecting that even the greatest of all men have died (III.1037–8). By contrast, he never mentions Hesiod, conventionally regarded as the originator of didactic verse, or Aratus, the best didactic poet of more recent centuries, at all. It is an eloquent silence.

In reality, Epicurus sat quietly in his garden at Athens, writing and contemplating, but his Roman disciple represents him as a kind of epic hero. Homer's Odysseus was a traveller, who 'saw the cities of many men and knew their mind' (*Odyssey* 1.3); Lucretius, for his part, turns his master into an Odysseus of the imagination, who has roamed the immeasurable universe in mind and soul. He is also a warrior of the spirit, like Achilles perhaps or a victorious Roman general, a man who has conquered super-stition and brought home in triumph the knowledge of the truth. Lucretius' diction is often lofty and sometimes old-fashioned, not because he looks back sentimentally to the past – he is in fact singularly free from that nostalgia for the good old days which infects so many Roman writers – but because he wants to evoke an epic grandeur. This grandeur is not merely decorative: it is part of the argument. The opponents of Epicureanism commonly treated it as a dull, drab creed; Lucretius' assertion is that, rightly apprehended, it is beautiful, majestic and inspiring. In this poem the medium really is the message.

Lucretius thus hugely extends the scope of didactic poetry, but this enlarging ambition carries him down as well as up. In places he is comic, satirical and partly colloquial. Some of the refutation of other philosophers in the first book is knockabout stuff. The discourses on death and on love which conclude the third and fourth books are diatribes in the ancient sense of the word – that is, not invectives, but rough, vigorous arguments of a quasi-popular kind, sometimes robustly humorous or cheerfully abusive. Parts of the diatribe on love are perhaps not too seriously meant, but the diatribe on death is central to his purpose, and here he combines satiric edge with declamatory grandeur to achieve a tone which is unique in Latin literature – or would be, except that it reappears in a few of the satires of Juvenal, a century and a half later. Lucretius could write with exquisite delicacy when he chose, but his most characteristic tone is dense, energetic and even a little cumbrous. Sometimes he is ostentatiously prosaic. He does not need to speak of 'Anaxagoras's homoeomery' (I.830), let alone repeat this lumpish Greek polysyllable; elsewhere he happily invents his own technical vocabulary, and he could easily have done so here too. But he is gripped by 'the fascination of what's difficult', to borrow Yeats's phrase, and where he is awkward or craggy, he is so with a purpose, because he is dramatizing his struggle with the recalcitrance of his material. He seizes dull or prosy language and forces it to become poetry, catching it up into an epic sweep and splendour. And this too is part of his message: he glorifies the factual, the ordinary and the everyday as no poet had before.

Many of Lucretius' readers have been struck by his vehemence; some have been surprised to find this quality in a man whose philosophy advocated an imperturbable calm; and a few have thought that he protests too much, revealing by his very assertiveness some inner doubt or insecurity: a nineteenth-century French scholar spoke of 'the anti-Lucretius in Lucretius' (M. Patin, *Études sur la poésie latine* (Paris, 1868), p. 117). This notion may have been encouraged by the story that Lucretius was sent mad by a love-potion and wrote his masterpiece between bouts of insanity. This tale has the merit of having inspired Tennyson's splendid poem *Lucretius* (1868), but its source is centuries later than the real Lucretius, and it should be disbelieved. The idea that the poet exposes his uncertainty despite himself is indeed misguided, but it does draw attention to one of his methods. He likes to show the emotional force of false beliefs, only to override them with Epicurean truth. When we look up at the stars in heaven, he says, it is hard indeed not to believe that there are divine intelligences in them (V.1204–10); but such belief is error. Many readers have been moved by the lament of the unphilosophic in the third book, warning that death will deprive a man of his dear wife and children, and those readers include two poets, for the passage is poignantly echoed in Virgil's *Georgics* and in Thomas Gray's *Elegy Written in a Country Churchyard* (1751). And although there is an element of satire in his tone, Lucretius does take the risk of making the lamentation genuinely affecting: the detail of the children running ahead to snatch the first kiss is so sweetly observed and the silent tenderness that touches the heart is so lovingly evoked that they are bound to move us (III.896–7). He affirms the strength of his faith by demonstrating that it can overcome the temptations that might assail it: he creates something authentically eloquent, and then in plain words rebuts the error, coolly crushing the beauty that he himself has made.

His sense of detail is indeed one of his virtues. In one respect Epicurus' philosophy, although rigidly materialist, was better suited to poetic treatment than a more exalted or transcendent doctrine would have been, because it insisted that the truth was to be attained through an accurate examination of the world around us. Lucretius notices the ring on a finger rubbed thin underneath, the hands of bronze statues worn smooth by the hands of passers-by, the stone hollowed out by the drip of water (I.311–18). He observes how children deliberately make themselves dizzy, and remembers how it felt (IV.400–403). He describes dreams as they really are, and not as literature represented them, recording the falling dream, the discontinuities within dreams and the way that dreams rework the experiences of waking life (IV.1020–23, 962–83, 818–22). He captures the fraction of a second between the starting-gates being thrown open and the horses bursting out (II.263–5). He studies the iridescence of a dove's neck, trying out colour words and the names of jewels in an effort to convey the exact reality of what he sees (II.801–4). He watches the hue changing as a dyed cloth is pulled into shreds (II.828–30); he enjoys the ripple of colours across the audience in a theatre as the awning flaps above them on a sunny day (IV.75–80).

These closely observed particularities are attractive in themselves, and they contribute to Lucretius' case for the atomic basis of all matter, but they also have a moral dimension. He looks at a shallow puddle lying between the stones that pave a street and sees that the reflections in it appear as deep as the sky is high above him (IV.414–20). He thus celebrates the ordinary world by making familiar sights appear both strange and wonderful. Elsewhere he invites us to contemplate the bright, clear colour of the sky, the planets, the moon and the sun in its splendour. What could be more astonishing, he declares, if we were seeing these things for the first time; but as it is, people are so jaded that they do not look up at the shining spaces of heaven (II.1030–39). The blight of modern life, he declares in yet another place, is a restless boredom, equally dissatisfied whether in town or country, at home or abroad (III.1060–67). By contrast, early mankind used to smile and laugh, because everything was new and marvellous (V.1403–4). Part of Lucretius' mission is to refresh experience: he wants not only to make us use our eyes but to give them back their innocence.

For all his feeling for detail, Lucretius also delights in immensity; no poet has relished infinity more than he. Much of his quality, indeed, resides in his power of combination – that is to say, in an ability to create a distinctive amalgam out of diverse and even opposite elements: loftiness and colloquialism, poetry and philosophy, the smallest particles of matter and the totality of the universe, the didactic and the epic modes. His success was of a peculiarly rare kind, for there are only two didactic poems which stand in the first rank of the world's masterpieces, *The Nature of Things* and Virgil's *Georgics*, written only a generation later. And that is not an accident.

The literary critics of antiquity commonly divided poetry up into genres: epic, lyric, elegy, comedy, tragedy, and so on. The Greeks mastered these genres, and the Romans sought to emulate them, but they often fell short of their models, and in only one genre, didactic, did they decisively outclass the Greek achievement. Through moral and intellectual intensity and (as Cicero saw) by possessing both craftsmanship and natural genius, Lucretius found possibilities in the didactic form that no one had found before, and thus, through his influence on Virgil, he was to affect the course of Western literature as a whole. Virgil began as a miniaturist, with the *Eclogues*, a collection of short, exquisite and elusive pieces. However, Lucretius' example taught him that it was still possible to write on the grand scale and that there was one genre, didactic, in which a Roman poet could hope to beat the Greeks at their own game. Unlike Lucretius, Virgil openly imitates Hesiod and Aratus; like Lucretius, and unlike his Greek predecessors, he infuses a moral and spiritual vision into the didactic form, in a profound exploration of man's relation to land, to nature and to his country. Lacking what one might call moral originality, most didactic poems, however skilful, are limited to formal ingenuity and descriptive charm, and are liable to look like more or less artificial literary exercises. Lucretius and Virgil have been perhaps the only two didactic poets who were fully able to overcome these limitations.

The Nature of Things showed Virgil the way to the Georgics, and the Georgics showed him that he could attempt an even longer poem, his epic Aeneid. But there is a deeper influence of Lucretius upon the Aeneid to be considered. Lucretius put the idea of salvation into the 'great poem' and Virgil then put it into the heroic poem. Virgil's Aeneas is a man who is on a quest to found a city, and the poem studies how human beings may hope for happiness and security. Lucretius finds safety within the individual, in the citadel of the soul; the Aeneid finds it in human institutions, in city and land, in tradition, law and custom. But the study of salvation would henceforward be recurrent in the 'great poem', in Dante's Divine Comedy (early fourteenth century), for example, and in Milton's Paradise Lost (1667). Milton, indeed, admired Lucretius greatly, and he may have the Roman poet consciously in mind as he inverts his generic play: whereas The Nature of Things is a didactic poem presented in heroic manner, Paradise Lost is a heroic poem presented as didactic, written to 'assert eternal providence, And justify the ways of God to men' (I.25–6). Lucretius would have deplored Milton's theology, but he would have appreciated his passion to teach and to persuade.

For all his vehemence, there is a reserve within Lucretius. He is proud and secret; if the shadowy Memmius who is named twice early in the poem (I.26, 42), and very sporadically addressed thereafter, was his patron, he gets little of the flattery that a patron was accustomed to receive. Other didactic poets disclose something of themselves: Hesiod reveals in his *Works and Days* that he has quarrelled with his brother, and that he has never been to sea, except for a very short crossing to an island and back; Virgil tells us in the *Georgics* about where he was born and where he is living now, gives some indication of his age and recalls an old gardener whom he once knew in the far south of Italy. By contrast, although few people read Lucretius without feeling his strong individuality, we learn nothing about the author himself. Like a voice on the radio, the voice may be very distinctive, but we do not know where it is coming from. That passionate reticence is part of his power: we are left, as he evidently intended that we should be left, with the poem, and the poem alone.

Richard Jenkyns

Further Reading

Classen, C. J., 'Poetry and Rhetoric in Lucretius', *Transactions of the American Philological Association* 99 (1968), pp. 7–118

Clay, Diskin, *Lucretius and Epicurus* (1983)

Dudley, D. R. (ed.), *Lucretius* (1965)

Fowler, Peta, 'Lucretian Conclusions', in *Classical Closure*, ed. D. Roberts, F. Dunn and D. Fowler (1997), ch. 6

Furley, David J., 'Lucretius the Epicurean on the history of man', in *Lucrèce*, vol. 24 (1978), ch. 1

Gale, Monica R., *Myth and Poetry in Lucretius* (1994)

Jenkyns, Richard, *Virgil's Experience*, Part 3, 'Lucretius' (1998)

Kenney, E. J., 'Doctus Lucretius', *Mnemosyne* 23 (1970), pp. 366–92

Sedley, David, *Lucretius and the Transformation of Greek Wisdom* (1998)

Segal, Charles, *Lucretius on Death and Anxiety* (1990)

Sykes Davies, Hugh, 'Notes on Lucretius', *Criterion* 11 (1931–2), pp. 25–42

Wallach, Barbara Price, *Lucretius and the Diatribe against the Fear of Death* (1976)

West, David, *The Imagery and Poetry of Lucretius* (1969)

The articles by Classen, Furley, Kenney and Sykes Davies are reprinted in C. J. Classen (ed.), *Probleme der Lukrezforschung* (Hildesheim, 1986).

A Note on the Text and Translation

It might seem crazy in modern times to render 7,400-odd lines of Latin poetry on physics and philosophy into rhyming fourteeners. I had no such plan when I set out. But in deciding, a decade ago, for a lark, to try my hand at some passages of Lucretius, I stumbled on heptameter as roomy enough to embrace the Latin dactylic hexameter in an almost line-by-line translation. Fashioning couplets (and the occasional triplet) out of the original had something of the pleasure of a crossword puzzle, though one to which there were no guaranteed answers. When I was commissioned to complete the poem, it seemed natural to continue as I had begun.

In the past two hundred years or so, Lucretius has gone from being scientifically prescient to outdated in the realm of atomic theory – we no longer read him for science (though, for instance, in his proto-Darwinian discussion of the evolution of life and his arguments against 'intelligent design' (IV.823 ff.), not to mention his warning against the potential evils of religion (I.80–101), he remains strangely topical). The surprising and wonderful thing about the *De Rerum Natura* is that it is a *poem*; this should be clear to the reader immediately, even in a translation. What more obvious way to convey poetry in English than rhyme? Rhyme is not the essence of our poetry, but it is I think the honey of it.

The standard metre nowadays (where there is one) for translation of classical verse is an unrhymed line of roughly five beats, distantly descended from the blank verse the Earl of Surrey invented for translating sections of Virgil in the 1540s. But fourteeners have their own pedigree, and with their kinship to ballad metre seem well suited to narrative, and thus epic, poetry in English. Fourteeners were widely used in the sixteenth century by English translators of Latin tragedy and epic, who seem to have considered it the native equivalent of the classical hexameter. Both Arthur Golding in his translation of Ovid's *Metamorphoses* (1567) – variously exalted by Ezra Pound as 'the most beautiful poem in the language' and decried by C. S. Lewis for its 'ugly fourteeners' – and George Chapman in his 1611 translation of the *Iliad* found the metre amenable to their purposes. Later, Alexander Pope's smooth heroic couplets would become the new standard in classical translation. Yet it was the native, rough-hewn vigour of the fourteeners, not Pope's Augustan polish, that fired Keats's appreciation for Homer ('On First Looking into Chapman's Homer'). And perhaps the fourteener's old-fashioned rhythm and ring will get across something of the archaic flavour of Lucretius' Latin without resorting to a fusty pose of Thees and Thous.

Lucretius complained of a paucity of scientific and philosophical vocabulary in Latin (he has to 'coin much new / Terminology', I.137–8); English is rich in it. I have not shied away from using current scientific terms where they are most direct, even if technically anachronistic. The translation is peppered with anachronisms, sometimes to make something more immediate to the modern reader (popping balloons rather than animal bladders, bullets rather than leaden missiles, glass mirrors rather than polished metal ones), sometimes because I couldn't resist ('news of things hot off the press' for 'the hot news of things', IV.703). And there are quite a few anachronistic literary nods, as if Lucretius were well versed in the English poets. I hope this gives something of the allusive texture of the poem that would have been felt by his ancient Roman audience steeped in the Greek poets and philosophers. After all, as T. S. Eliot reminds us, for the individual, the whole of Western literature has 'a simultaneous existence and composes a simultaneous order'.

Some might find the overt religious language in passages surprising, especially where it feels proto-Christian. But Lucretius has the religious fervour of a convert when it comes to Epicureanism. Indeed, it has been suggested that early Christian communities were influenced by Epicurean ones. (Epicureanism has the strange quality of being both pervasive and invisible in its influence. People often mention, for instance, that Thomas Jefferson's ringing phrase 'Life, Liberty and the Pursuit of Happiness' is altered from John Locke's triad of Life, Liberty and Property, but rarely note that both are drawing on Epicurus. Jefferson wrote in a letter to William Short, 31 October 1819: 'I too am an Epicurean.')

To avoid an excess of notes cluttering the text, proper names and places are defined in the Glossary. Also in the interest of simplicity, I have kept notes on the text and line-order to a minimum, and have not indicated where I have occasionally flipped line order for English syntax or have conflated Latin lines. (I probably average a little over nine lines to every ten in the original. Thus the line numbers, by tens, in the text are geared to the Latin of Cyril Bailey's Oxford Classical Text (1947); while in my translation, there might occasionally be eight, nine or even eleven lines between 'decades'.) Writing in the epic tradition of Homer, Lucretius occasionally repeats phrases, lines and even passages verbatim. Within the constraints of a rhymed translation, this effect was not always possible to replicate, and so I sometimes make use of variation where Lucretius uses repetition. I have not generally noted where I have taken minor alternative readings to Bailey. Lacunae I have marked with ellipses – [. . .] – or, where I have supplied the probable sense of a lacuna, usually in accordance with Bailey, I have done so in square brackets.

It is strange to think that the fate of this major poem seems to have rested (precariously) with a single fourth- or fifth-century-AD manuscript, known as the Archetype, now lost to us, and a descendant, also lost, the source in turn for all other copies. Incomplete ends of several consecutive lines, therefore, may mean a page of that work was smudged or damaged; a lengthy lacuna might mean an entire leaf (there were 26 lines to the page) was torn out. There is suddenly a glimpse back through the centuries, real parchment suffering the indignities of time or water or fire or mice; mortal hands, straining eyes. Though the poem's vision of a material universe with no Providence at the helm must have seemed threatening to many a monk and copyist, nonetheless there must have been others who saw that (to quote Charles Darwin) 'There is grandeur in this view of life.'

I was surprised and delighted to learn, after I embarked on this project, that in all likelihood the first person to english the entire poem (in the 1640s or 1650s) was a woman, the memoirist Lucy Hutchinson. She had no previous translation to guide her, nor was her text in the same state we have it today, yet she rendered this difficult poem with admirable accuracy into lively rhyming couplets, in order 'to understand things I heard so much discourse of at second hand'. Her description of the circumstances under which she worked, with children underfoot and while she was engaged in domestic tasks, will ring a bell with many female scholars and poets:

> I turnd it into English in a roome where my children practizd the severall qualities they were taught with their Tutors, and I numbred the syllables of my translation by the threds of the canvas I wrought in, and sett them downe with a pen and inke that stood by me. (ed. Hugh de Quehen, 1996)

As I had a baby son during the last two years of this project, who proceeded to sprout up into an energetic toddler, it was encouraging to think that the person who had first blazed this trail was a busy mother.

I, of course, have had access to texts, editions, commentaries and translations that Lucy Hutchinson could only have dreamed of. The lucid prose of R. E. Latham's *On the Nature of the Universe* (1951) was of help in an overview of passages which are difficult to digest – from vagueness or technical subject matter – at a line-by-line level. Rolfe Humphries' brisk, blank verse translation *The Way Things Are* (1969) often spurred me to greater vigour and concision. While the line numbers in my translation correspond on the whole with the OCT, the text I lived with most was the eminently portable, revised Loeb (Rouse, 1975), with its wealth of annotations. I also have on my bookshelf the edition of Leonard and Smith (1942).

My early drafts were in a looser, accentual line with slanter rhymes, but several readers challenged me to strive for a stricter measure. My skill and my understanding of the poem improved over time, but there are necessarily palimpsests of these earlier efforts peeping through. It occurs to me that if I could start it all

again, I would do much better justice to Lucretius' masterpiece –
but that, like Lucretius himself, I might never complete it.

A. E. Stallings

Some Works Consulted

The New Princeton Handbook of Poetic Terms, ed. T. V. F.
   Brogan (1994)

Titi Lucreti Cari De rerum natura libri sex, ed. with Prolegomena,
   Critical Apparatus, Translation and Commentary by Cyril
   Bailey, 3 vols. (1947)

*Lucretius: The Way Things Are. The De Rerum Natura of Titus
   Lucretius Carus*, trans. Rolfe Humphries (1969)

Lucretius: De rerum natura, Book III, ed. E. J. Kenney (1971;
   1991)

Lucretius: On the Nature of the Universe, trans. R. E. Latham;
   rev. John Godwin (Penguin Classics, 1951; 1994)

T. Lucreti Cari, De rerum natura, libri sex, ed. William Ellery
   Leonard and Stanley Barney Smith (1942)

T. Lucreti Cari De rerum natura libri sex, ed. H. A. J. Munro
   (1900)

Lucy Hutchinson's Translation of Lucretius' De Rerum Natura,
   ed. Hugh de Quehen (1996)

Lucretius: De rerum natura, trans. W. H. D. Rouse; rev. Martin
   Ferguson Smith (1975; 1992)

Acknowledgements

This project would never have come to fruition without the suggestions, encouragement and assistance of numerous people. Robert Harris at the University of Georgia sparked my interest in Virgil and first suggested the beauty and strangeness of Lucretius to me, quoting the lines about flocks of stars grazing their way across the night sky. Rick LaFleur at UGA first encouraged me as an aspiring poet to have a go at verse translation, and my first verse translation, of a poem of Catullus, was written for his graduate workshop. Richard Jenkyns directed my Virgilian studies at Oxford and guided my attention not only to the huge influence of Lucretius ('felix qui potuit rerum cognoscere causas'), but to Lucretius' special greatness as a poet. It was Richard Jenkyns who, on seeing my early attempt at Book I, suggested to Peter Carson I might complete the entire poem. Monica Gale took great pains over early drafts of my translation, pointing out misunderstandings, inaccuracies and infelicities. If I have persisted in error, or created new problems in revision, it is my own folly. Philip Thibodeau, with a Latinist's eye and a poet's ear, also corrected my reading of some passages, and the poet and translator Dick Davis generously scoured an entire draft for poetic missteps. Some of these, of course, I have retained out of stubbornness. (Rare is the poet who can murder all her darlings, however misbegotten.) My husband, John Psaropoulos, showed enormous patience in looking over every line, sometimes in half a dozen (maddeningly and) minutely differing variations. Peter Carson has been a source of bracing comment, feedback and, most importantly, encouragement throughout. Monica Schmoller, at the copyediting stage, gave many invaluable suggestions and helped me to polish rough edges. I am grateful also to Hawthornden Castle International Writers' Retreat in Scotland for a month 'of peace in decent ease' to work on revisions and to the American School of Classical Studies in Athens for access to the Blegen Library. I would also like to thank *The Hudson Review*, *The New Criterion* and *Two Lines*, journals in which brief passages of the translation originally appeared.

I am sorry my father, William M. Stallings, did not live to see this book. He shared with Lucretius a passion for the truth and a love for the natural world – even a fondness for hunting metaphors. I dedicate this to his memory.

THE NATURE OF THINGS

BOOK I

MATTER AND VOID

Life-stirring Venus,[1] Mother of Aeneas and of Rome,
Pleasure[2] of men and gods, you make all things beneath the
  dome
Of sliding constellations teem, you throng the fruited earth
And the ship-freighted sea – for every species comes to birth
Conceived through you, and rises forth and gazes on the light.
The winds flee from you, Goddess, your arrival puts to flight
The clouds of heaven. For you, the crafty earth contrives sweet
  flowers,
For you, the oceans laugh, the skies grow peaceful after showers,
Awash with light. For soon as morning wears the face of spring,   10
And the West Wind is free and freshens, warm and quickening,
The airy tribe of birds, O Holy One, is first to start
Heralding your approach, struck with your power through the
  heart;[3]
Then beasts, the wild and tame alike, go romping over the lush
Pastureland and swim across the rivers' headlong rush,
So eagerly does each pant after you, so do they heed,
Caught in the chains of love, and follow you wherever you lead.
All through the seas and mountains, torrents, leafy-roofed
  abodes
Of birds, and greening meadows, your delicious yearning goads
The breast of every creature, and you urge all things you find
Lustily to get new generations of their kind.   20
Because alone you steer the nature of things upon its course,
And nothing can arise without you on light's shining shores,
And nothing glad or lovely can be fashioned, I invite
You Goddess, stand beside me, be my partner as I write

The Nature of Things, these verses I am striving to set down
For Memmius, my friend, your favourite, whom you would
    crown
With every honour and with everlasting accolades –
More reason to endow my words with grace that never fades.

Meanwhile, Holy One, both on dry land and on the deep,
30 Make the mad machinery of war drift off to sleep.
For only you can favour mortal men with peace, since Mars,[4]
Mighty in Arms, who oversees the wicked works of wars,
Conquered by Love's everlasting wound, so often lies
Upon your lap, and gazing upwards, feasts his greedy eyes
On love, his mouth agape at you, Famed Goddess, as he tips
Back his shapely neck, his breath hovering at your lips.[5]
And as he leans upon your holy body, and you reach
Your arms around him, Lady, sweet-talk him with honeyed
    speech,
40 Pleading for a quiet peace for Romans – this I ask,
For I cannot with easy mind perform my chosen task,
Nor can the noble scion of the Memmii fail to heed
The call to duty, when our land is in her hour of need.[6]

For godhead by its nature must enjoy eternal life
In utmost peace, removed from us and far from mortal strife,
Apart from any suffering, apart from any danger,
Powerful of itself, not needing us, and both a stranger
To our attempts to win it over and untouched by anger.[7]

50 For what's to come, open your ears, apply keen intellect
Far from cares, to true philosophy, lest you reject
Out of hand the gifts that I've assembled for your sake
Before you've fully grasped them. For I now begin to make
My discourse on the lofty law of gods and heaven above,
And shall reveal the building blocks all things are fashioned of,
Nature's prime particles, from which she nourishes and grows
All things, and into which once more she makes them
    decompose.

We term them in philosophy, according to our needs,
Matter, atoms, generative bodies, elements and seeds,   60
And first-beginnings since it is from these that all proceeds.

When human life lay on the ground obscenely, in full view,
Prostrate, crushed beneath the weight of Superstition, who
Stretched down her head from heaven's realms and with her
    ghastly gaze
Loomed over mortal men, the first among them who dared raise
His human eyes to her was Greek,[8] the first man to withstand
    her.
Neither the myths of gods, nor lightning bolts, nor threatening
    thunder
Of heaven hindered him but, rather, all the more they fired
His mind's courage, so that he was the first man who desired   70
To break the close-barred gates of Nature down. The vital force
Of his intelligence prevailed, and he advanced his course
Far past the blazing bulwarks of the world, and roamed the
    whole
Immeasurable Cosmos in his mind and in his soul.
In triumph he returns to us, and brings us back this prize:
To know what things can come about, and what cannot arise,
And what law limits the power of each, with deep-set boundary
    stone.
Therefore it is the turn of Superstition to lie prone,
Trod underfoot, while by his victory we reach the heavens.

One thing I am concerned about: you might, as you commence   80
Philosophy, decide you see impiety therein,
And that the path you enter is the avenue to sin.
More often, on the contrary, it is *Religion*[9] breeds
Wickedness and that has given rise to wrongful deeds,
As when the leaders of the Greeks, those peerless peers, defiled
The Virgin's altar with the blood of Agamemnon's child,
Iphigenia.[10] As soon as they bound the fillet[11] round her hair
So that its ends streamed down her cheeks, the girl became
    aware

That waiting at the temple for her there would be no groom –
Instead she saw her father with a countenance of gloom
90 Attended by the priests who kept the blade well hid. The sight
Of people shedding tears to see her froze her tongue with fright.
She sank to the ground upon her knees. It did not mean a thing
For the princess now, that she had been the first to give the king
The name of *Father*. No, for shaking, the poor girl was carried
By the hands of men up to the altar, not that she be married
With solemn ceremony, to the accompanying strain
Of loud-sung bridal hymns, but as a maiden, pure of stain,
To be impurely slaughtered, at the age when she should wed,
Sorrowful sacrifice slain at her father's hand instead.
100 All this for fair and favourable winds to sail the fleet along! –
So potent was Religion in persuading to do wrong.[12]

Sooner or later, you will seek to break away from me,
Won over by doomsayer-prophets. They can, certainly,
Conjure up for you enough of nightmares to capsize
Life's order, and churn all your fortunes with anxieties.
No wonder. For if men saw that there was an end in sight
To trials and tribulations, they would find the power to fight
Against the superstitions and the threats of priests. But now
110 They have no power to resist, no way to reason how,
For after death there looms the dread of punishment for the
     whole
Of eternity, since we don't know the nature of the soul:
Is the soul born? Or does it enter us at our first breath?
And does it die with us, and is it broken down at death?
Or does it haunt the murk of Orcus and his vasty halls?
Does it slither by some magic into other animals? –
So Ennius declares, the first among us to bring down
From fair Mount Helicon an evergreen and leafy crown,
Thus making his name famous throughout all of Italy;
Yet even so, he sets forth in his deathless poetry
120 That realms of Acheron exist – there really is a Hell –
And there we neither in the flesh nor in the spirit dwell
But, rather, something wraithlike of us lingers, wan and weird.
And it is from these same infernal regions there appeared

The shade of never-fading Homer, who, the poet sings,
Began to shed salt tears and to unfold the Nature of Things.

Therefore we must consider well celestial happenings,
And by what principle the sun and moon run on their courses,
And all phenomena upon the earth, and governing forces.
And then especially, we must nose into, with sharp wits,   130
What makes up the soul, and what the nature of it is;
What do we meet when we're awake, delirious with fever,
That terrifies the mind, or when we're sepulchred in slumber,
So that we think we see and hear such persons, face to face,
Who have encountered death, and whose bones lie in Earth's
    embrace?

Nor does it fail me that discoveries – obscure and dark –
Of Greeks are difficult to shed much light on with the spark
Of Latin poetry, chiefly since I must coin much new
Terminology, because of our tongue's dearth and due
To the novelty of subject matter. And yet to this end
Your excellence and my sweet hope to win you as a friend   140
Persuade me to tackle any task and take up any toil,
And in the still, small hours, make me burn the midnight oil,
As I seek the right words and the right poetry to light
Brilliant lanterns for your mind, so that at last you might
Peer deep into the recesses of things once recondite.

This dread, these shadows of the mind, must thus be swept away
Not by rays of the sun nor by the brilliant beams of day,
But by observing Nature and her laws. And this will lay
The warp out for us – her first principle: *that nothing's brought*
Forth by any supernatural power out of naught.   150
For certainly all men are in the clutches of a dread –
Beholding many things take place in heaven overhead
Or here on earth whose causes they can't fathom, they assign
The explanation for these happenings to powers divine.
Nothing can be made from nothing – once we see that's so,
Already we are on the way to what we want to know:

What can things be fashioned from? And how is it, without
The machinations of the gods, all things can come about?

For if things were created out of nothing, any breed
160 Could be born from any other; nothing would require a seed.
People could pop out of the sea, the scaly tribes arise
Out of the earth, and wingèd birds could hatch right from the
        skies.
Born willy-nilly, every animal, both wild and tame,
Would inhabit cultivated land and wilderness the same.
The same tree would not always grow the same fruit – what
        might bear
An apple one time, might, the next, produce a quince or pear.
Since there would be no generating particles, then neither
Would certain things arise from only a certain kind of mother.
But since in fact each species rises from specific seeds,
Each thing springs from the source that has the matter that it
        needs,
170 The primary particles, and comes into the boundaries
Of light, and that's the reason every thing cannot give rise
To every other thing, because there is a separate power
In distinct things. For why else do we see the roses flower
In spring, grain ripen in the heat, and under autumn's sway
The grapes pour forth, if not because each in its proper day,
When the right seeds come together, opens out into its birth
When seasonable weather is at hand and when the teeming earth
Brings tender growth forth safely onto the shores of light? It's
        clear
180 That if things came from nothing, they would suddenly appear
At random intervals and at the wrong time of the year;
For there would be no basic particles of generation
To be hindered from a fruitful meeting in a hostile season.
Nor, in turn, would things need time to grow, for congregation
Of basic particles, if things could grow up out of nothing,
For babes would shoot up into youths in a flash, and groves
        would spring
Suddenly leaping from the earth. But clearly this is not
What happens. All things grow, little by little, as they ought,

From a certain seed, preserving their own species as they go,   190
So each thing needs its own kind of material to grow.
Consider that without a certain season of rain, the earth
Could not put forth her gladdening fruits. Nor could creatures
    give birth
To young or stay alive deprived of their food. It makes more
    sense,
Therefore, to think that many things have common elements,
As words share letters, rather than assume that anything can
Exist without them.

                   Again, why cannot Nature make a man
So large that he could wade across the deep to other lands,   200
Mighty enough to wrench apart great mountains with his hands
And outlive generations, unless everything consists
Of certain matter, and this matter limits what exists?
We must, therefore, confess that *nothing can emerge from*
    nothing,
But every thing created needs a seed from which to cast
Into the gentle breezes of the air. Another thing,
Because we see tilth is more fruitful than untended land
And gives back better harvests cultivated by our hand,
It's plain that in the ground the elements of things are rife,   210
Since turning rich clods with the plough, we stir them into life.
If no such elements existed, then you'd see the soil
Would grow a richer crop all on its own, without our toil.
Add that *Nature does not render anything to naught,*
But she instead reduces everything that she has wrought
Back to its elemental particles again. For say
That any thing, in all its parts, were subject to decay –
Then snatched of a sudden from our sight each thing would pass
    away,
For there would be no need of any force to make a chink
Between component parts and to unfasten link from link.   220
But since each thing is made of atoms, those seeds that abide
Forever, until it meets a force that's able to divide
It with a blow, or that can enter it where there is void,
Then Nature won't allow it to be openly destroyed.

Besides if Age consumes all the material outright
Of everything the lapse of time has taken from our sight,
Then out of what does Venus bring the creatures, breed by
    breed,
Back to the light of life? And what does the crafty earth feed
Them on and make them grow, each kind according to its need?
What feeds the sources of the ocean – both those fountains
    found
Below its surface and the rivers flowing overground?
230 Or the aether the stars graze upon? For all that *can* decay,
Devoured by the ages, should by now have passed away.
But if, in all the span of days gone by, something has stayed,
The stuff that makes the universe, with which it is re-made,
These particles perforce are indestructible. Therefore
Nothing can be reduced to nothing, as I said before.

Again, one cause would send all things wholesale to their demise
If they weren't knit together, loosely or tightly, from the ties
240 Of everlasting matter. For the mere tap of a feather
Would be sufficient to destroy such things not put together
From particles of eternal substance; there would be no call
For a certain force to fray the bonds of their material.
But as it is, since elements are of eternal stuff
Linked with bonds of different strengths, unless a strong enough
Force encounter it, a thing stays safely as it was.
Therefore nothing turns to nothing. All things decompose
Back to the elemental particles from which they rose.

250 And finally, when raindrops are cast down from Father Sky
Into the lap of Mother Earth, they vanish from the eye,
But gleaming crops rise up, and trees put forth green leaves and
    shoots,
And the trees begin to grow, and weigh their branches down
    with fruits,
And so we in our turn are nourished, and so the wild brutes –
Hence we see happy cities all abloom with girls and boys,
And the trills of fledgling birds fill up the leafing woods with
    noise,

And herds and flocks, made sluggish with their fat, lay down
    their bulk
In rich pastures, their heavy udders oozing with white milk,
And lambs go frolicking across young grass on wobbly legs –   260
Their new-born noggins tipsy on milk drunk straight from the
    kegs!
Thus things that seem to perish utterly, do not. See how
Nature refashions one thing from another, and won't allow
A birth unless it's midwived by another's death.

                        Come now,
Since I have taught that nothing's made of nothing, and once
    brought
Into existence cannot be recalled again to naught,
Just in case you start to think this theory is a lie
Because these atoms can't be made out by the naked eye,
You yourself have to admit that there are particles
Which *are* but which cannot be seen. First, take the force of
    squalls   270
That whip up, lash the ocean, founder sea-going ships and
    scatter
The clouds, or sweeping hurricanes that scour the plains and
    batter
The mountaintops with forest-snapping blasts, strewing their
    path
With massive trees, so fiercely does the wind roar in its wrath.
Thus clearly there are particles of wind you cannot spy
That sweep the ocean and the land and clouds up in the sky,
Snatching them in a sudden twister. And these bodies flow
And deal their devastation just like water – for although   280
Water is soft, yet when a stream is swelled by heavy rain
That avalanches in an instant down a towering mountain,
It crashes broken branches and whole trees together. Not
Even stalwart bridges can withstand the sudden onslaught,
So mighty is the swirling mass of rain upon its course,
So strong the river hitting the sturdy piles with all its force.
The flood wreaks havoc with a roar and rolls huge stones below
Its rush, dashing away all obstacles standing in its flow.

290 Therefore it follows gusts of wind are borne along as well,
     Just like a powerful river, when they come sweeping down
             pell-mell
     Driving all things before them – blow by blow, battering down –
     Or snatching things up into the rapid swirling of a cyclone.
     And so I say, again and again, that wind is made of matter,
     For though invisible, it acts in the same way as water,
     Which clearly is a substance.

                                      Moreover, we sense various smells
     Although we never see the odour wafting towards our nostrils.
300 We cannot see the scalding heat, or see the cold; likewise,
     We are not in the habit of noting voices with our eyes.
     But what acts on our senses is material, inasmuch
     As without body, nothing is tangible, nor can it touch.
     Moreover, clothing hung out by a breaker-beaten shore
     Grows damp, but if you spread it in the sun, it dries once more.
     Yet how the moisture came and went, you cannot see at all –
     And so the water must evaporate in drops so small
310 They escape detection by our eyes. Year after circling year,
     The ring upon a finger thins from inside out with wear.
     The steady drip of water causes stone to hollow and yield.
     The curving iron of the ploughshare fritters in the field
     By imperceptible degrees. The cobbles of the street
     We see are polished smooth by now from throngs of passing
             feet.
     And at the city gates, right hands of statues made of brass
     Are worn away by touches of the greeting hands that pass.
     And thus we see things dwindle by their being rubbed away –
320 But what is lost at any given moment, we can't say
     Because our stingy sense of sight will never let us see.
     Lastly, whatever Days and Nature add on gradually
     To things, that makes them grow at a moderate pace, cannot be
             seen
     No matter how you stare and squint, your sight however keen.
     Neither are you able to perceive what disappears
     At any given time, when things grow old and waste with years

Or rocks are gnawed away with sea-brine. Thus these things are
    done
By Nature using bodies that are visible to none.

Yet the universe is not one solid mass, all tightly packed:
There's also emptiness in things, which is a useful fact   330
In many matters, one that will not let you go astray,
Perplexed about the universe and doubting what I say.
For if there were no emptiness, nothing could move; since it's
The property of matter to obstruct and to resist,
And matter would be everywhere at all times. So I say
Nothing could move forward because nothing would give way.
But as it is, we notice that before our very eyes
Many things are moving various ways throughout the skies
Above our heads, and on dry land, and in the briny ocean.   340
Sans void, these would not only lack for agitated motion
But existence altogether. They could no way come to pass
With all things at a total standstill, chock-a-block with mass.

Why, even solid things are not as dense as they appear:
Everything is riddled with emptiness, as I'll make clear.
Take rock – through walls of stony caves, a clammy moisture
    seeps,
And all around, the place with many drops of water weeps.
Food is dispersed throughout the frames of animals. And fruits   350
Are put forth in good time by a growing tree, as it distributes
Nutrients through trunk and branches, from the deepest roots.
Voices pass through walls and fly through rooms across closed
    doors.
Stiffening chills creep to the marrow. But if there were no pores,
No openings that particles were able to pass through,
None of this would happen. Lastly, why do we see two
Objects of the same size differ in their weight? Instead,
If a ball of yarn contains the same amount of mass as lead,   360
Then they should weigh the same, since mass's property is to
    press
All downwards, while the property of void is weightlessness.

So that which is of lighter weight but seems the same in size
Reveals without a doubt it has more void within. Likewise,
An object of the same size that is heavier, must contain
More matter in it and much less of emptiness, it's plain.
Clearly, therefore, what we're tracking with keen-scented wits –
What we call void – is tangled up with things, and must exist!

370 Here I must run ahead, warn you what some wrongly declare,
Lest you be detoured from the truth. For certain people swear
That waters yield to scaly fishes nosing through and make
Liquid channels open up, since fish leave in their wake
Room for the yielding waves to flow together, so however
Full the universe, things can change places with each other.[13]
I'll have you know this line of reasoning can have no base.
For where can scaly fishes swim if water won't give place?
380 And where can water flow back into if the fish can't budge?
So matter is either deprived of movement, or else we must judge
Void is enmeshed in things, and is where movement gets its start.

Lastly, if two objects bump and then leap far apart,
Air must rush in to occupy the vacuum. Furthermore,
No matter with what swiftness the swirling currents flow
      together,
Even so they cannot fill the empty space at once,
But flow first to the nearest part, and then the next, and thence
390 They fill up the entire space. If anyone, perchance,
Thinks this happens right when the two objects spring away
Because the air condensed itself, then he has gone astray.
For in that case a vacuum's made that wasn't there before
While void that pre-existed isn't empty any more,
And even if air *could* compress that way, it's doubtful whether
Sans void, it could condense and draw its particles together.

Therefore, drag your feet and make excuses how you will –
Yet emptiness exists in things – you must confess it still.
400 Though I can scrape together many proofs to make my case,
These few, faint footprints are sufficient; you yourself can chase

The others down, with your keen nose for reason. As a pack
Of hounds, once it has stumbled on the sure scent of the track,
Can sniff out mountain-ranging quarry from its thickety lair,
So you can find one clue after the next in this affair,
And wriggle into every den and drag the truth from there
Lurking deep inside. But if instead you are remiss
And shrink a little from our task, then I can promise this,   410
Memmius, upon my word: my honeyed tongue shall pour
Such bountiful and copious fountains from my mind's rich store,
I fear that sluggish age will creep upon us with its chill
And loose the locks of life before I've had a chance to fill
Your ears with all the flood of evidence that I can use
Here in my verse on any given subject that you choose!

To pick up the thread where I left off: the universe's nature
Consists, in essence, of two different things: for there is matter,
And there is void, in which the particles of matter move   420
Hither and thither. The senses that men have in common prove,
In and of themselves, that matter *is*. Unless we place
Our firm faith in sensation, we shall have nothing to base
Conclusions on concerning what lies hidden from our view,
Nor could our reasoning confirm that anything is true.
Then further if there were no place or space, that which we call
Void, then particles would not have anywhere at all
In which to be or move in any direction to or fro –
Which I have demonstrated to you just a while ago.

Furthermore, there isn't anything that you could count   430
Entirely distinct from void or body, to amount
To a sort of third nature. For anything that *is* must be,
By definition, *something*. If it can affect the touch,
However faintly, then it adds its mass – however much
Or little – to the Sum of Things, if it exists at all.
And yet if, on the other hand, it is intangible,
And offers no resistance, so that anything that moves
Can pass through any part of it, without a doubt that proves
That it is void. Besides, whatever exists, will either do   440
Something, or it is itself, by other things, done *to*,

Or it will be where things exist and where events take place.
But unless something is empty and vacant, it cannot offer space;
Neither can any thing, sans body, be acted on or act.
Therefore, other than void and substance, there cannot be, in
     fact,
Any third nature existing in its own right – neither one
That falls at any time within the range of our perception,
Nor one that we can figure out by means of the mind's reason.

For you will find that everything for which we have a name
450 Is either a quality of the two,[14] or consequence of the same.
A quality is what, without obliterating shock,
Can never be separated and removed: as weight to rock,
As heat to flame, wet to water, the ability to touch
To every substance, intangibility to void. But such
As slavery, penury and riches, freedom, war and peace,
Whatever comes and goes while natures stay unchanging, these
We rightly tend to term as 'consequences' or 'events'.
Nor does Time exist in its own right. But there's a sense
460 Derived from things themselves as to what's happened in the
     past,
And what is here and now, and what will come about at last.
No one perceives Time in and of itself, you must attest,
As something apart from things at motion and from things at
     rest.

Further, when people talk about the rape of Helen or
Make mention of the trouncing of the Trojan tribes in war[15]
As *real*, beware they do not make us see it in the light
That these events exist as entities in their own right,
Seeing the generations these events befell have all
Long ago been swept away by ages past recall.
For any incident of history can be referred
470 To as a 'consequence' – of men or of where it occurred.
And had there been no stuff of things, and nowhere to take
     place,
Love of Helen's beauty never could have fanned the blaze

That smouldered under Alexander's heart and set alight
The luminary battles in the war's ferocious fight.
No Trojan-duping wooden horse, carrying in her womb
The night-mare birth of Greeks, would have sent Troy to
    burning doom.
So you can see that done deeds have no separate entity
The way that matter does, nor can these deeds be said to *be*
In the selfsame way that void *exists*. But we can rightly class   480
Them consequences of matter and the place they came to pass.

Further, substances consist either of a single kind
Of atom, or are made of compounds, elements combined.
But the atoms of things are such, there is no power that can
    snuff
Them out, for they prevail at last with stoutness of their stuff.
And yet it's hard to believe that anything of solid mass
Exists in all the universe. For bolts of lightning pass
Through walls of houses, just as noises and as voices do.
Iron glows white in fire, stones with fierce heat are split in two.   490
Warmth loosens up and melts down the frigidity of gold,
And the 'ice' of bronze is thawed in fire. Heat and biting cold
Trickle through silver – something anybody understands
Who's ever held a silver cup of vintage in his hands
When sparkling water's poured in.[16] So it seems a certain bet
That there is nothing solid in the universe. And yet
True reason and the Nature of Things say otherwise. Lend an
    ear
For just a few more verses until I can make it clear
That there are things made up of solid stuff that lasts forever,   500
Which I teach are the atoms, seeds of things that make the sum,
The basic elements the universe is fashioned from.

First, since we've found that Nature has a double-sided face –
Contrary aspects – matter, and the void where things take
    place,
Each one has to exist in its own right, and unalloyed;
Because wherever there is emptiness, which we call void,

There matter isn't, and whatever place is occupied
By matter, in that selfsame spot no vacuum can abide.
510 Thus atoms have no void contained within them and are sound.

Furthermore, since there is void contained within compound
Substances, then there must be some solid matter round
The void. Nor by true reason are you able to explain
How something can hide void within itself and can contain
Emptiness, unless you grant that whatever's there to hold
It in is solid. And only a union of matter can enfold
Void within it. Thus matter, made of solid stuff, can stay,
Lasting forever, while other things dissolve and fade away.

520 Again, were there no emptiness, then everything would stand
A solid mass. Were there no body, on the other hand,
To fill a space and take it up, then everything there is
Would consist entirely of vacancy and emptiness.
Thus body is distinct from void – that's indisputable –
But alternates with it, since the universe is not all full,
Neither is it empty. Therefore there are certain bodies
To set apart the filled-up places from the vacancies.
These atoms cannot be undone, struck by external blows,
Neither can they be penetrated so they decompose,
530 Nor can they be assailed in any way and made to crack,
Which I already demonstrated to you not far back.

For we see that without void nothing can be crushed or dashed
To pieces or crumbled into bits or cut in twain or smashed,
Nor can it take in moisture, nor admit the piercing flame,
Nor let in trickling cold, the instruments that are to blame
For everything's undoing. The more void something holds, the
    more
That the attacks of these will leave it shaken to the core.
If atoms are solid, therefore, and without void, as I've taught,
540 Then they must be eternal. And what's more, if they were not,
Everything already would have met with its demise,
And any objects that we see here, now, before our eyes

Would have been born again from nothingness. But since I've
    taught
That nothing's made of nothing, and that nothing can be
    brought
To nothingness once it is made, then there must be first bodies
Made of stuff that lasts forever – atoms – and it is these
That everything is broken down to in its final hour
So there is a supply of matter on hand to re-power
The world. These atoms thus are pure and solid through and
    through:
How else could they survive infinite time to make things new?   550

Furthermore, if Nature had not set a boundary
To the breaking down of things, then matter would already be
So reduced by all the splintering of ages past
That naught could be conceived at a given time, nor could it
    last
To reach the prime of life; because it's faster, it is plain,
For something to be broken down than to be made again.
And thus what had been broken down in all the time before –
The rough and tumble of Yesterdays – nothing could restore
In all the Tomorrows left to come. But in reality   560
There is a limit fixed to breaking down, because we see
Each thing is made anew, and each thing at its given time,
According to its breed, attains the flower of its prime.

Though atoms are composed of absolutely solid stuff,
How they in turn make up soft substances is easy enough
To explain – such substances as air and water, earth and fire[17] –
And how it is all things are done, by what force they transpire,
Once we include void in the mix of things. On the other hand,
Were atoms soft, there'd be no explanation to understand   570
What composes sturdy flint and iron, for in that case
All Nature would profoundly lack a cornerstone at base.
Thus there are particles of pure solidity and might,
And when these are combined, the more they're packed together
    tight,

The more the things that they create are tough and hard. Let's
     say,
For the sake of argument, no bounds were set for the decay
Of particles – yet even so it's clear some have prevailed
To make up things – bodies that have never been assailed
580 By any danger whatsoever. Yet, once we agree
They're breakable, they can't have lasted through eternity
Harried throughout the aeons by innumerable blows.

But since in fact a limit's set to how much each thing grows,
According to its kind, since there's a limit to its span
Of life, and since what each thing cannot do and what it can
Is governed by the laws of Nature, since each species stays
True to type, so true each different kind of bird displays
590 Down through the generations the marks belonging to its name,
Each also must contain material that stays the same.[18]
For if the atoms could be overthrown and made to change,
What could or could not come to be would be unsure, the range
Of possibilities would have no deep-set boundary stone,
Nor could the generations so often take after their own,
Repeating their parents' nature, behaviour, mode of life and
     motion.

Then furthermore, since when we peer at objects, there must be
600 An ultimate, smallest point which is the smallest we can see,
So also in things, there is a smallest point *beneath* our sight,
And this contains *no parts*, being of a stuff so slight,
It is the smallest stuff of all. And it can never start
To exist as something separate, because it's always part
Of something else, primal and indivisible. The way
Matter is composed is from such parts in tight array.
And since they can't exist alone, then they must closely cling
To the atom, and cannot be torn away by anything.
Atoms therefore are a pure and simple solidness,
610 Made of those smallest parts cohering tightly in a mass.
Atoms aren't assemblages made out of parts; they get
Their might from their eternal singleness. Nature won't let

Anything be wrenched from them, or any dwindlings,
But keeps them in one piece preserved to be the seeds of things.

Besides, unless there is a smallest part, however small
A thing may be, it must have infinite parts, since after all
Half of a half of anything can still again be cut
In two, and on and on *ad infinitum*. And then what
Will be the difference between the tiniest speck of matter
And all the universe? There won't be any whatsoever!   620
For even if the Sum of Things is infinite, the amount
Of smaller parts in the tiniest speck is likewise past all count.

But since sound reasoning cries out in protest that this can't
Be true, and the mind refuses to believe it, you must grant
That there are things that have no parts, and that these things
      consist
Of nature's smallest material. And since such things exist,
Atoms, you must grant, are solid, and last forever as well.
If Nature the Creator did habitually compel
All things to decompose to smallest parts, she would not then
Have the power, as she does, to rebuild things again.   630
A substance that's not fashioned out of parts cannot partake
Of qualities that matter must possess if it's to make
Anything – the various links, encounters, motions, mass
And blows – the means by which all things that happen come to
      pass.

Therefore those who think the basic element is fire,
And it is out of this material that the entire
Universe is made, have clearly wandered far astray
From the path of true reason. The first warrior in the fray
Is Heraclitus,[19] a leading light for the murkiness of his style
More so among the shallow, than those Greeks who find
      worthwhile   640
The search for truth. For idiots admire things all the more
When they discern them hidden in tangled words, and set great
      store

In anything that tickles the ear, in phrases dyed a shade
Of purple. I should like to know, if everything is made
Out of pure fire and out of fire alone, how can it be
The universe embraces such a rich diversity?
Because it would not be of any use for burning flame
To rarefy or thicken if its parts possessed the same
Qualities as fire as a whole. For a more dense
650 Array of particles would make the burning more intense,
While having particles more scattered and spread out, in turn
Would only make it that the fire would less fiercely burn.
Those are the only possibilities you can entertain –
So how can density and rarity of fire explain
The wide variety the world is able to contain?

Then this: if they[20] would just admit that there is emptiness
Mixed up in things, then fire could rarefy or could condense.
And yet because their muse perceives some dangerous ground
     ahead,
And shies away from leaving unmixed void in things, instead,
They veer from the steep drops, only to wander off and stray
660 From the true path. Nor do they see that if we take away
The void from things, then all would be condensed into one mass,
Nothing could send something shooting from itself, not as
A burning fire does, emitting warmth and shedding light,
So fire cannot be made of parts crammed up together tight.

But if by chance they think there is some way to rearrange
The fires so that their sparks are quenched, to make their
     substance change –
(That is, if they don't shrink from things so far-fetched and so
     strange!) –
Clearly heat would be entirely annihilated
And everything from nothingness would have to be created.
670 Anything so changed it isn't bounded any more
By its own limits, dies right then to what it was before.
So something must survive unscathed, something has to stay,
Or everything must utterly to nothingness decay
And out of naught the store of things be born and thrive again.

Now since there *do* exist particular bodies which maintain
The same nature forever, and which, when they're rearranged,
Or some are added or taken away, their qualities are changed,
And the substance that they make transforms itself, without a
    doubt
The possibility that these are fire can be *ruled out.*
For if you took away or added particles of flame   680
Or altered their formation, the result would be the same:
They all would still possess the properties of fire. Rather,
I think that there are certain bodies by whose union, order,
Place, shape, motion make up fire, but when they rearrange
Their structure, then the nature of what they create will change,
Neither do they resemble fire, nor any other such
That sends out particles that touch upon our sense of touch.

Further, to claim that everything is fire, as that man claims,[21]   690
And that the only real thing in the universe is flames,
Is to be off your head. It's from the senses that he makes
His argument, then takes up arms against them, and so shakes
The very bedrock of belief. For how does he perceive
That fire of his? By trusting those same senses that 'deceive'
Us when perceiving other things, things which are no less clear.
This line of reasoning strikes me as lunatic and queer!
For what will be our reference point? To what can we appeal?
What is more certain than our senses to tell false from real?   700
And besides, why should every other thing be taken away
And fire be the only thing that is allowed to stay?
Why not abolish fire instead, let something else remain?
What does it matter? – both ideas are equally insane.

Therefore those who've theorized that only fire exists,
And it is out of fire alone the universe consists,
And those who've thought the building blocks for making things
    are air,
And those who've thought that water fashions all things
    everywhere,
Or else that earth produces everything, and earth can change
Itself to any substance – we can see these people range   710

Far afield from truth. And add to those the ones who pair
The 'elements' of things – the folk who yoke together air
With fire, and earth with water, and those people who suppose
That everything out of this elemental foursome grows:
That is, out of wind and rain and earth and fire.

                                        Of these,
Standing in the forefront is the great Empedocles,
Born in Acragas, on Sicily, with its three sides
Of winding coastline washed by the grey-green Ionian tides
And showered with salt-spray. A stretch of the sea separates
720 The island from the shores of Italy with narrow straits
And rushing currents. Here the treacherous Charybdis lies,
Here Aetna rumbles threats to hurl its lightning to the skies
Once more, and vomit forth the angry flames out of its maw.
And although many look upon this wondered land with awe,
And say it is a sight to see, a land of plenty, strong
In men, yet even so, of all the wonders that belong
To this isle, none is holier, and none brings greater fame,
730 None is more wonderful, and none more precious than his
      name!
And since the ringing poems of his god-like mind proclaim
His great illuminations in a voice so loud and clear,
He scarcely seems to be a man born in the mortal sphere.

Nevertheless this man, and those of whom I spoke before
(Though they, by infinite degrees, are his inferior),
Despite the excellent discoveries that they impart,
Inspired divinely from the holy of holies of the heart
With more of sanctity and with a great deal more of truth
Than have the utterances of the Sibyl saying sooth
Throned on Apollo's tripod, breathing in the laurel's fume,
740 Yet when it comes to fundamentals, there they meet their doom.
These men were giants; when they stumble, they have far to fall:
First, because they allow for motion with no void at all.
And while they will permit things to be soft and rarefied –
Animals and plants, air, sun, rain, earth – they don't provide

For any void to be mingled in their bodies. Second, because
They set no limit to the division of bodies, and no pause
To their decay, and utterly deny that there can be
A smallest particle; though there's a point in things we see
That is the ultimate, the least our senses bring to light,   750
And so you may likewise deduce, in things *beneath* our sight,
There is an ultimate point which is the smallest part of all.
Third, since the 'elements' they've picked are soft – things we
      would call
'Compounds', stuffs completely subject to decay and rot –
Everything by now would have dissolved again to naught,
And out of naught the store of things be born and thrive anew
(Two notions that by now you know are very far from true).
And fourth, these elements in many ways are foes and strive,
The bane of one another; if they meet, they don't survive,
Or else they leap apart, as when we see a tempest gather,   760
And out of its clouds the wind and rain and bolts of lightning
      scatter.

In short, if the whole universe were fashioned from these four,
And everything dissolved back to these elements once more,
Why are they called the *elements* of *things*? Why not instead
Consider *things* their *elements*, and turn it on its head?
For they are born of one another, ever changing hue
And changing their entire natures with one another too.
But if by chance you think that particles of earth and fire   770
And windy air and flowing water so combine together
That their natures do not alter in their unions, you will find
That nothing can be fashioned from such elements combined,
No animal, nor anything inanimate, like a tree,
But every item added to this jumbled assembly
Will display its own nature: air will be discerned in earth,
And fire be seen in water. For when elements give birth
To things, they should bring with them natures hidden and
      unknown
That don't stand out or thwart or hinder others' being shown,   780
So what they've made displays a nature that is all its own.

What's more, they think that everything originates in the sky
With the sky's fires, claiming that the fires can modify
Themselves to air, and that the air can be transformed to rain,
That rain turns into earth, and all returns from earth again,
Turning first to water, thence to air and last to fire,
Nor do these ever cease from changing into one another,
Wandering from sky to earth, from earth to fiery stars.
790 That elements should act like this is something reason bars:
There must be something never changes – something has to stay,
So all things do not utterly to nothingness decay.
For anything so changed it isn't bounded any more
By its own limits, dies right then to what it was before.
So since the things just mentioned change, there must be
     something other
Of which they are created, and which cannot ever alter,
Or else you'd find all things return to utter nothingness.
Wouldn't you rather grant that there are bodies that possess
Such a nature that if they should happen to make fire,
800 By adding or subtracting, changing their motions or their
     structure,
The same could create air, and all could change into each other?

'But the evidence is clear,' you say, 'the facts openly show
It's up into the air and out of the earth that all things grow
And are nourished. Unless the season at the proper time allowed
The rains to fall, so trees could shiver in the melting cloud,
Unless the sun for his part cherished them, and gave them
     warmth,
Not crops nor trees nor creatures would be able to spring forth.'

Fair enough. And without solid nourishment and fresh
810 Water to sustain us we should quickly lose our flesh,
And life would fray from every bone and sinew. For it's plain
That it is certain substances that feed us and sustain,
While other things are fed on other substances – no doubt
Since many basic particles common to things are mixed about
In different ways. So different things need different sustenance.
And how these atoms are arranged makes all the difference –

Their position and formations, and what moves they give and
    take
From one another, for the selfsame atoms go to make
The heavens and the sea, the land, the rivers and the sun,   820
The same make crops, trees, animals – but by their combination
In different ways with different elements move differently.
Furthermore, all through these very lines of mine, you see
Many letters that are shared by many words – and yet
You must confess that words and lines from this one alphabet
Have sundry sounds and meanings. Letters only have to change
Their order to accomplish all of this – and still the range
Of possibilities with atoms is greater. That is why
They can create the universe's rich variety.

Let's turn to Anaxagoras's homoeomery[22] –   830
Due to the dearth of our Mother tongue, I use the term of the
    Greeks,
Though the thing itself is easy to explain. First, when he speaks
Of homoeomery in things, it's clear that he supposes
That bones are made of itsy-bitsy bones, and what composes
Flesh is teensy gobs of flesh, what joins up and becomes
Blood is bloblets of blood, and gold is made of golden crumbs,
Earth is an amalgamation of little earthen grains,   840
Fire of little fires, water of waters. And he maintains
This system goes for everything else. And yet he will not let
Any emptiness exist in things, and will not set
A limit to the division of bodies. He's on a par, therefore,
In both of these wrong views with those we've spoken of before.
Add that his elements are feeble – *if* things that partake
Of exactly the same properties as substances they make
Can be considered *elements* – and suffer and pass away
The same. Nothing reins them back from ruin and decay.   850
For which one of these 'elements' will hold up underneath
The terrific strain, and flee Death from between Doom's very
    teeth?
Will it be fire or water or air? Will it be blood or bone?
Which one of these will be able to last? In my opinion, none,

When all is fundamentally as subject to demise
As are those things that manifestly vanish from our eyes,
Defeated by some force. I call as witness what I've taught
Earlier to demonstrate that nothing can be brought
Back into nothingness, nor be created out of naught.
Besides, because food nourishes the flesh and makes it grow,
860 Blood and veins and bones [are made of alien stuff], we know.
Or if they say that nutriment's a compound that contains
Small particles of gore and so on – sinew, bones and veins –
It follows that all food, all solid food and liquid too,
Consists of stuffs that are unlike itself: of bone and sinew,
Blood mixed with pus. Whatever substances spring from the
        ground,
If they are inherent in the earth, earth must be found
To consist of those same foreign stuffs to which the earth gives
        rise –
Change the subject and the selfsame string of words applies.
870 If fire and smoke and ashes all are lurking inside wood,
Then wood consists of stuffs unlike itself, it's understood,
Of those same alien substances that it emits. And so
Whatever bodies the earth nourishes so that they grow
[. . .]²³
And here is where the loophole for evasion lies – the one
That Anaxagoras exploits in claiming all is hidden
And intermixed in everything else, yet what is manifest
Is the main ingredient, more visible than all the rest,
880 Placed in the fore. The reasoning behind this isn't sound,
For in that case it would be natural that, when grain is ground,
Crushed beneath the oppression of the millstone, there would
        show
Some sign of blood, or something of those substances that grow
In our bodies, or when we grind stone on stone, that blood
        should flow.
Likewise grass and water with sweet droplets ought to ooze,
Savouring of milk from udders of the fleecy ewes.
And then, when clods of soil are crumbled, you should see
        revealed
The different flora, crops and vegetation there concealed,

Scattered and lurking in minute amounts within the ground.   890
And finally, inside of broken wood, there should be found
Small particles of ash and smoke and fire concealed from view.
But since experience teaches us that none of this is true,
We can be sure that things are not all jumbled in this fashion
With one another, but that there must be certain seeds in
     common
To many things, mingled in many ways, hid from our sight.

'And yet,' you say, 'it often happens on a mountain height
That towering trees, whose topmost limbs are right next to each
     other,
Are made, by violent southern winds, to scrape their boughs
     together
Until the flower of flame is kindled and they catch on fire.'   900
This is true without a doubt. Yet fire is not innate
In wood. No, wood has many seeds of heat which congregate
By friction and which set the forests blazing. If inside
Of wood were lurking full-fledged flames, the fires could not
     hide
For a moment. They'd devour all forests, burn up all the trees.
You see now what I said when talking of Empedocles –
That it makes all the difference in the world in what position
And with what bonds the atoms are held in place, what kind of
     motion
They give and take from one another. By just a slight change   910
In the structure of its elements, the wood can rearrange
To fire. Even the letters of the words are much the same,
Shuffled around, if we refer to them as 'maple' and 'flame'.[24]

And lastly if you think whatever you see before your eyes
Cannot come about unless the elements that comprise
The whole possess the properties of that which they devise,
You stumble, and your very elements come tumbling after:
Next you'll have them shake with giggles, quivering with
     laughter,
And sprinkling their faces and their cheeks with salty tears![25]   920

Now pay attention to what follows and prick up your ears –
Nor does it escape me how obscure this all appears.
But the goad of hoped-for glory strikes my spirit to inspire,
And at the selfsame moment smites my heart with sweet desire
For Muses, stirring up my thoughts. My mind abuzz, I blaze
New trails across their mountain haunts, among untrodden
    ways.
I thrill to come upon untasted springs and slake my thirst.
I joy to pluck strange flowers for a glorious wreath, the first
930 Whose brow the Muses ever crowned with blossoms from this
    spot.
Why? Because I teach great truths, and set out to unknot
The mind from the tight strictures of religion, and I write
Of so darkling a subject in a poetry so bright,
Nor is my method to no purpose – doctors do as much;
Consider a physician with a child who will not sip
A disgusting dose of wormwood: first, he coats the goblet's lip
All round with honey's sweet blond stickiness, that way to lure
940 Gullible youth to taste it, and to drain the bitter cure,
The child's duped but not cheated – rather, put back in the
    pink –
That's what I do. Since those who've never tasted of it think
This philosophy's a bitter pill to swallow, and the throng
Recoils, I wished to coat this physic in mellifluous song,
To kiss it, as it were, with the sweet honey of the Muse.
That is the purpose of my poetry, as you peruse
My lines, to try to keep your mind's attention, while you start
950 To understand the framework at the universe's heart.

But since I've taught that atoms are as solid as can be,
And flit, unconquered, endlessly throughout eternity,
Come now, let us unfurl if there is any upper bound
To their sum, and also as regards that void that we have found
Exists – place or space where each thing comes to pass – let's
    see
Whether its extent is bounded fundamentally,
Or else it opens measureless and fathomlessly deep.

The universe must therefore have no limits in its sweep
In all directions, for if it did, then it would have a bound,
And if it has a boundary, then something must surround   960
It from without, so that the eye can follow only so
Far and no farther. And since we must confess that there is no
Thing *beyond* the universe, then it can have no border,
And stretches limitless and without end. Whatever quarter
You stand in makes no difference. Whatever place you are,
It stretches out in all directions infinitely far.
But let's say for a moment Space *were* limited. Pretend
That someone with a spear goes running to the very end
And hurls the whizzing missile. Does the spinning spear then go   970
Flying afar along the trajectory of the mighty throw,
Or do you think that something thwarts it, standing in its path?
You must confess just one of these is true – you can't have
      both –
Yet each shuts your escape hatch and compels you to confess,
Whichever one you choose, the universe is limitless.
For whether there is something there to thwart the missile's
      flight
So it falls short of its target, or it passes on outside,
It was not launched from any boundary. I'll persevere:
Wherever you set the furthest brink, I'll ask about the spear.   980
The result is that no last frontier can ever stay in place –
For possible flight forever pushes back the edge of Space.

Furthermore, if all of Space were limited and bound
With definite boundaries on every side the whole way round,
By this time, sinking underneath its solid weight, the store
Of matter would have sifted from every side down to the floor
Of the universe. Nor could anything at all be done
Under the tent of heaven; there would be no shining sun,
Nor any heaven either. Instead, there would only lie a vast   990
Heap of matter, sunken down through endless ages past.
But as things stand, the basic particles are not allowed
Any respite, since the universe is not endowed
With any fundament at all, no place where they might flow
Down to and settle. But always, every thing is on the go

In every corner, and atoms are supplied and ever flit,
Stirred up ceaselessly, out of the bottomless Infinite.

Lastly, we see before our eyes that one thing bounds another.
Air is a wall between the hills, and mountains hem the air,
1000 Land shuts in the sea, and ocean borders every shore,
But nothing immures the universe, since there is nothing more
Beyond it. Therefore, Space is such, and its abyss so deep,
That brilliant bolts of lightning are not able in their sweep
To travel it from end to end, not even should they fly
Along the endless course of Time. Nor are they able by
Their travelling to lessen the distance that they have to go,
So widely does the huge supply of Space unfold, with no
Borders, out in all directions. Nature won't permit
The Sum of Things to ring itself with any kind of limit,
Because she makes it such that matter has to be surrounded
1010 By emptiness, and emptiness must be by matter bounded,
Making sure, by alternation, the cosmos has no end.
But either one, unbounded by the other, would extend
Even so, all by itself, throughout infinity.
[. . .
Matter is also infinite], or neither land nor sea
Nor gleaming quarters of the sky, nor men, nor holy race
Of gods would be able to endure an hour's slender space,
For matter, with its gatherings disbanded and destroyed,
Would have drifted apart as lonely atoms through the void.
Or rather, atoms would never have come together to create
1020 Anything, since, scattered, they could never congregate.

For certainly the elements of things do not collect
And order their formations by their cunning *intellect*,
Nor are their motions something they agree on or propose;
But being myriad and many-mingled, plagued by blows
And buffeted throughout the universe for all time past,
By trying every motion and combination, they at last
Fell into the present form in which this universe appears.
Indeed, this form, which has endured through many aeons of
    years,

Once set in proper motions, makes the rivers' generous streams   1030
Slake the thirsty sea, and makes the earth, warmed by the beams
Of the sun, renew its brood, and tribes of living things arise
And thrive, and quickens celestial fires that glide along the skies.
None of this would be possible if there were no supply
Of matter that could rise up from the Infinite, whereby
All that has been lost can in due season be replaced.
Just as any animal deprived of food will waste
And lose its flesh, so every thing must likewise fade away
Once its supply of matter is cut off or turned astray.   1040
Nor can external blows slamming from every side maintain
The integrity of a world composed of atoms. They can rain
Down thick and fast, and slow the dissolution for a space
Until more matter is supplied to come and take the place
Of what is lost, and yet they must bounce back sometimes,
    and so
Allow the atoms opportunity and room to go,
That they can fly away from their combinations. And that's why
I say again and again that there must be a huge supply
Of atoms. Even to keep the aforesaid rain of blows supplied   1050
Would require an infinite amount of matter on every side.

There is one theory, Memmius, to keep your distance from –
That everything is pressing towards the centre of the Sum,
As certain people claim.[26] That's how the world can stand its
    ground
Without a hail of outside blows to hem it in all round,
Nor can the top or bottom parts be loosened even a little,
Since every part of it is always pressing towards the middle
(If you believe that anything upon itself can rest),
And all those weights that lie below the earth are
    upward-pressed,
Resting upside-down beneath the earth, as things appear,
Reflected in the water, to us on the surface here.   1060
They likewise claim that creatures ramble topsy-turvy there
And can no more fall down into the caverns of the air
Than we can float up to the heavens in spontaneous flight,
And that when they behold the sun, we see the stars of night,

And that the seasons of the heavens come to us by turns,
And when it's night for one, then for the other daylight burns.
But all of this is just a fairy tale, pipe dreams of fools
Who base their reasoning upon a faulty set of rules.
1070 How can there be a centre to what's infinitely vast?
And even if there *were*, then why should anything stand fast?
Why shouldn't it be driven far away instead? That space
That we call void, central or not central, must give place
To all things that have mass, wherever their movements
       navigate,
Nor is there any spot where bodies lose the force of weight
So they can stand still in the void. Nor can the void provide
1080 Support for anything – it yearns by nature to subside.
And thus the cosmos cannot be contained by such a course,
Held together in the grip of centripetal force.

Besides, they also think there are exceptions to this trend,
And only particles that are of earth or water tend
To the centre – the waters of the sea, the torrents that descend
The mountainsides, and anything that's bodied out of clay,
While claiming air and fire at the same time drift away
From the centre, and that's why the firmament is all ablaze
With constellations, and that is how the burning sun can graze
1090 Across the blue of heaven, since all of the heat that flees
The earth collects there. Yet, how could the topmost twigs of
       trees
Leaf, unless each drew up food from earth, little by little?
[How can this be if everything of earth tends to the middle?[27]
  . . .]
Lest in a flash the ramparts of the cosmos be destroyed,
Scattering, as flying flames do, through the vasty void,
And taking the rest with them. The sky's provinces of thunder
Would crash down from above, the earth would slip away from
       under
Our very feet, and midst their tangled ruin and decay,
Their elements set free, heaven and earth would pass away
Into the fathomless abyss. And all that would be left,
1110 In the blink of an eye, would be invisible atoms and bereft,

Deserted Vacancy. Whatever place you shall assume
Particles first disappear, there gapes the gate of Doom
For things. All matter, jumbled, will go tumbling out this door.

And thus you will gain knowledge, guided by a little labour,
For one thing will illuminate the next, and blinding night
Won't steal your way; all secrets will be opened to your sight,
One truth illuminate another, as light kindles light.

BOOK II
THE DANCE OF ATOMS

How sweet it is to watch from dry land when the storm-winds
   roil
A mighty ocean's waters, and see another's bitter toil –
Not because you relish someone else's misery –
Rather, it's sweet to know from what misfortunes you are free.
Pleasant it is even to behold contests of war
Drawn up on the battlefield, when you are in no danger.
But there is nothing sweeter than to dwell in towers that rise
On high, serene and fortified with teachings of the wise,
From which you may peer down upon the others as they stray
This way and that, seeking the path of life, losing their way:
The skirmishing of wits, the scramble for renown, the fight,
Each striving harder than the next, and struggling day and
   night,
To climb atop a heap of riches and lay claim to might.
O miserable minds of men! O hearts that cannot see!
Beset by such great dangers and in such obscurity
You spend your little lot of life! Don't you know it's plain
That all your nature yelps for is a body free from pain,
And, to enjoy pleasure, a mind removed from fear and care?
And so we see the body's needs are altogether spare –
Only the bare minimum to keep suffering at bay,
Yet which can furnish pleasures for us in a wide array.

Nature has no need of more, not golden figurines
Throughout the house, with lamps in their right hands to light
   the scenes

Of nightly feasts and revelry, nor does nature require
A palace[1] whose enamelled ceiling echoes to the lyre,
Glinting silver and gold, when people can as pleasantly pass
Their hours *al fresco*, sprawling out in groups on the soft grass
Beside a babbling brook, beneath a tall and shady tree,   30
Where they can merrily unwind, and practically for free –
Especially on spring days when the weather smiles serene
And when the season sprinkles flowers all across the green.

Nor do raging fevers any faster cease to burn
If you have fancy tapestries on which to toss and turn
And royal-purple sheets to wrap in, than if you are broke
And all you have to huddle under is a peasant's cloak.
Therefore since riches do not do the body any good –
And nor does rank nor royalty – it should be understood
That neither do they profit the mind – that is, unless, of
     course,
When you behold your legions swarm the Field of Mars[2] in
     force,   40
Waging a mock battle, well supplied in foot and horse,
All fitted out alike, one equal temper of hearts – or when
You watch your tossing fleet of ships spread far and wide[3] – that
     then
Your superstitious fears are frightened by this and depart
Trembling from your brain, and Dread of Death flees from your
     heart
Leaving it free and loosed from care. But if instead we see
These things are laughable, mere jokes, and in reality,
The cark and care of mortal men and all their hounding fears
Are not alarmed by clanking armour nor by savage spears,
But brazenly consort with kings and potentates, and bold,   50
Will neither bow to purple robes nor to the glint of gold,
Why doubt that reason alone can quench this terror with its
     spark,
Especially since life is one long labour in the dark?
And just as children shudder at everything in black of night,
So sometimes things we are afraid of in the broad daylight

Are only bugbears such as tots dread in a darkened room,
And therefore we must scatter this terror of the mind, this
     gloom,
60 Not by the illumination of the sun and his bright rays,
But by observing Nature's laws and looking on her face.

Now by what motion atoms come together to create
Various things, or how these things once formed can dissipate,
And by what force they are compelled, and what freedom of
     motion
They have to meander through the vasty void, I shall explain,
Just pay close attention. Clearly matter's not compressed
Into one heap, because we notice things becoming less,
And we perceive that, over time, everything ebbs and wanes,
70 And old age steals them from our sight, while yet the sum
     remains
Undiminished. This is because the particles that go
From one shrinking object cause another thing to grow,
Making the former shrivel up, while making the latter flower,
Never lingering. Thus the Sum of Things is every hour
Renewed, and thus, in order to thrive, all mortal creatures need
Each other. While some species are ascendant, some recede,
And generations are renewed again in a brief space,
Passing on life's torch, like relay runners in a race.[4]

80 If you should think these atoms have the power to stop and stay
At a standstill, and set new motions going in this way,
Then you have rambled far from reason and have gone astray.
Since atoms wander through a void, then they must either go
Carried along by their own weight or by a random blow
Struck from another atom, seeing that when atoms crash
Into one another, they bounce apart after the clash
(And no surprise, since they are hard and solid, and they lack
Anything behind them to obstruct their moving back).

All bodies of matter are in motion. To understand this best,
90 Remember that the atoms do not have a place to rest,

And there's no bottom to the universe, since Space does not
Have limits, but is endless. As I have already taught
And proved with reason irrefutable, it opens wide
And far in all directions, measureless on every side.
And therefore it is obvious no respite's ever given
To atoms through the fathomless void but, rather, they are
    driven
By sundry restless motions. After colliding, some will leap
Great intervals apart, while others harried by blows will keep
In a narrow space. Those atoms that are bound together tight,   100
When they collide with something, their recoil is only slight
Since they are tangled up in their own intricate formation:
Such are the particles that form the sturdy roots of stone,
And make up savage iron and other substances of this kind.
Of the other particles drifting through the vast deep, we find
A few leap far apart and bounce a long way back again,
Providing us with thin air and the shining of the sun.
And many more besides stray through the void, either out cast
From combinations, or which alliances could not hold fast   110
In harmonious motions.

                  There's a model, you should realize,
A paradigm of this that's dancing right before your eyes –
For look well when you let the sun peep in a shuttered room
Pouring forth the brilliance of its beams into the gloom,
And you'll see myriads of motes all moving many ways
Throughout the void and intermingling in the golden rays
As if in everlasting struggle, battling in troops,
Ceaselessly separating and regathering in groups.   120
From this you can imagine all the motions that take place
Among the atoms that are tossed about in empty space.
For to a certain extent, it's possible for us to trace
Greater things from trivial examples, and discern
In them the trail of knowledge. Another reason you should
    turn
Your attention to the motes that drift and tumble in the light:
Such turmoil means that there are secret motions, out of sight,

That lie concealed in matter. For you'll see the motes careen
130 Off course, and then bounce back again, by means of blows
          unseen,
Drifting now in this direction, now that, on every side.
You may be sure this starts with atoms; they are what provide
The base of this unrest. For atoms are moving on their own,
Then small formations of them, nearest them in scale, are
          thrown
Into agitation by unseen atomic blows,
And these strike slightly larger clusters, and on and on it goes –
A movement that begins on the atomic level, by slight
Degrees ascends until it is perceptible to our sight,
140 So that we can behold the dust-motes dancing in the sun,
Although the blows that move them can't be seen by anyone.

Now then, Memmius, it will not take you long to read
About the movement of the atoms, their amazing speed.
First of all, when dawn is sowing the earth with her fresh light,
And parti-coloured birds throughout the pathless groves take
          flight
And fill the glades with song and flood the air with melody,
The moment that the sun comes up, how instantaneously
Light saturates the world and mantles everything with gold
Is evident to everyone and obvious to behold.

150 But as the heat and cheerful light the sun sends here below
Does not come travelling through a vacuum, it is forced to slow
Down as it cleaves the waves of air. The bodies of heat do not
Move individually, but linked together in a knot,
So that at once they drag each other back, and are slowed, too,
By friction from without. But atoms, solid through and through,
Unities – though having parts – encounter no delay
When hurtling through the void with nothing standing in their
          way;
Every one towards a certain destination wends
160 Along the trajectory that their initial motion tends,
Certainly with greater nimbleness and at a pace
Much faster than the brightness of the sun, since they can race

Across a much, much wider stretch of distance in the same
Time that it takes the sun to flood the heavens with its flame.
[. . .]⁵
Nor to follow every single atom, one by one,
In order to perceive by what force every thing is done.
But certain people,⁶ ignorant of matter, are at odds
With this, and think it is impossible without the gods
For Nature to create the crops and alternate the seasons   170
In such convenient accordance with our human reasons,
And when they daydream it's for *our* sake that the gods arrayed
Everything in the universe, these men have grossly strayed
From reason's strait and narrow in every way. I might not know
That such a thing as atoms of matter existed – even so,
From the very workings of the skies above I would be bold
And claim – a deduction many other examples would uphold –
In no way was the universe made by the power of God   180
For *our* sake, when the universe stands so profoundly flawed.
I'll clarify that to you later. Meanwhile, I'll explain
All things about the motions of the atoms that remain.

This is the point I shall establish for you in due course
That *nothing physical can lift itself by its own force,*
Nor can it drift up on its own, nor is it able to rise.
Do not let fire fool you into thinking otherwise!
Yes, flames start upward, and they leap up higher as they grow,
And upward rise the gleaming crops and trees too, even though
However much they weigh, that weight is always tugging down.   190
And when a fire pounces on the rooftops of a town
And laps up beams and timbers with its flicking tongues of
        flame,
Don't think it hurls itself up on its own. It's just the same
With blood, when it is let out of the flesh and starts to pour,
And spurts up in a throbbing fountain, spattering with gore.
And haven't you seen with what force water spits out logs and
        lumber?
For the deeper that we shove them and the more we push them
        under,

Many of us pressing down with all our might and main,
That much more eagerly the water spews them out again,
So that the beams leap up out of the water with such strength
200 They jump into the open air by more than half their length.

Even so, there cannot really be a doubt, I think,
That all these things – as much as it is in their power – sink
Downward through an empty void. And that is why flames rise,
Because they are squeezed up by draughts of air, though their
      weight vies,
As much as it can, to drag them down. Don't you ever gaze
Up at the heavens and behold the torches there that blaze
In the night sky, and how they drag behind a fiery trail
Along whatever trajectory that Nature lets them sail?
And haven't you seen stars and other heavenly bodies plummet
210 To the earth? And take the sun, broadcasting heat down from
      the summit
Of heaven out in all directions, sowing the fields with light;
Thus heat as well tends towards the earth. And haven't you
      caught sight
Of lightning flitting slantwise through the thunderheads? – a
      flash
Now from this part, now from that – fires run around and
      smash
Out of the clouds. And often blazing bolts drop to the ground.

Another basic principle you need to have a sound
Understanding of: when bodies fall through empty space
Straight down, under their own weight, at a random time and
      place,
They swerve a little. Just enough of a swerve for you to call
220 It a change of course.[7] Unless inclined to swerve, all things
      would fall
Right through the deep abyss like drops of rain. There would be
      no
Collisions, and no atom would meet atom with a blow,
And Nature thus could not have fashioned anything, full stop.

But if anyone should chance to think that heavier bodies drop
Straight down through the void with greater speed, so as they go
They catch up with, and strike, the lighter particles below,
And thus give rise to fertile motions, he has gone astray,
Wandered far from the path of reason and has lost his way.
Why? Whatever falls through water or thin air, the rate   230
Of speed at which it falls must be related to its weight,
Because the substance of water and the nature of thin air
Do not resist all objects equally, but give way faster
To heavier objects, overcome, while on the other hand
Empty void cannot at any part or time withstand
Any object, but it must continually heed
Its nature and give way, so *all things fall at equal speed*,
Even though of differing weights, through the still void. And so
Heavier bodies will never strike the lighter ones below,   240
Nor by themselves will they initiate a blow that sets
The divers motions going out of which Nature begets
Creation. Thus, I repeat, the atoms have to swerve a little,
But only by the smallest possible degree, a tittle –
We do not want to look as though we thought things moved
     along
In sideways motion, when the Truth would come and prove us
     wrong!
For as far as you can see, weights falling from above can't veer
Sideways – that is something that is obvious and clear.
But do weights never waver by the slightest bit and stray
Out of their vertical path by just a little? Who can say?   250

Again, if every motion is connected, and we hold
New motions that arise, arise in due course from the old,
And atoms do *not* swerve a little and initiate
The kind of motion which in turn shatters the laws of fate,
But leave effect to follow cause inexorably forever,
Where does that freewill come from that exists in every creature
The world over? Where do we get that freewill, wrenched away
From the fates, by which we each proceed to follow pleasure's
     sway,

So that we swerve our motions not at a designated spot
260 And fixed time, but the very place we will it in our thought?
Without a doubt these motions have their beginning in the
        whims
Of each, and from that Will these motions trickle into the limbs.

When the starting gate swings open at the races, don't you see
How the horses' energy, champing at the bit, cannot burst free
As quickly as the mind itself desires? For the whole supply
Of matter in the flesh must be spurred on, with a great try
Throughout the frame, so it can follow the yearning of the mind.
270 And therefore motion has its impetus in Thought, we find,
First rising from a whim of spirit, then travelling all through
The flesh, and through the limbs. The same, however, is not true
When we lurch forward because we have received a mighty
        shove
From someone *else* – it's clear then our whole mass is made to
        move
And that our body's rushed ahead involuntarily
Until our freewill curbs it back throughout our limbs. You see,
Don't you, that even though a force outside them may propel
A crowd, sometimes stampeding them against their will,
        pell-mell,
280 Yet there is something in our chest can fight back and can stand
Against it, making the mass of matter turn at its command
Throughout our body, and when that mass is spurred ahead, can
        rein
It back into its place and settle it back down again.

Therefore, it's clear, the principle for atoms is the same:
There must be something *else* besides just weights and blows to
        blame
Their movement on, from which our power of freewill arises,
Because we know that nothing can emerge from nothingness.
290 (It's weight that stops all movement from resulting, you will
        find,
From blows – external pushes, as it were.) And why the mind

Is *not* bowed by necessity in everything it does
And forced to just endure it and to suffer is because
Of the slight swerve of the atoms, at a random time and place.

The store of matter never was more densely packed together,
Nor more scarcely scattered, for it does not increase ever,
Neither does matter perish, so the way atoms move *now*
Is just the way they must have moved in ages past, and how
They *shall* move in the time to come. And it is no surprise
That under the same conditions the same things tend to arise,   300
And all of these things thrive and grow and strengthen to the
    extent
That each of them has been allowed by Nature's covenant.
Nor can any power change the universe – there's no
Place beyond the universe to which matter could *go,*
No place outside the universe from which a new supply
Of matter could arise to burst inside the Sum thereby
Changing the whole Nature of Things, and altering the course
Of its motions.

                    And another thing that should not be a source
Of wonder, that although its atoms are in constant riot,
The universe itself seems to be standing still and quiet   310
Except for any object that displays its own motion.
The nature of these particles lies outside our perception,
And so, because the particles themselves are out of sight,
Their motions must escape our notice, too. It's only right,
Since even with the visible, great distance can disguise
Its motions, so that what we see is hard to recognize:
Often, a woolly flock of sheep upon a hillside crops
A verdant pasture all a-spangle with the dew's fresh drops,
And where the grass entices them, they wander one by one,   320
And the plump lambs gambol, and kick up their heels in fun;
But when we view them from afar, the distance blurs the
    scene
Until it's just a patch of white against a field of green.

Also, when mighty legions fill the rolling fields to clash
In games of war, there rises to the sky above a flash,
And the land, reflecting so much bronze, flickers all around.
The ground quakes underneath them and the mountainsides
    resound
With shouts and marching feet, and throw the echo to the stars.
And galloping in circles, suddenly a troop of horse
330 Surges down the field, and shakes the earth beneath their force.
Yet seen from a tall mountaintop, they would seem to remain
As though at a standstill, but a gleam upon a plain.

Next, let's turn to how the atoms vary in their kind,
And all the many different shapes of atoms that we find.
That's not to say that many do not share a shape – I claim
Only that they're not all universally the same.
It should be no surprise, since I have proved to you before
340 That there's no limit to the atoms' sum, that furthermore
There must be many shapes and forms collected in that store.

And then consider Mankind and its peoples, if you wish,
And all the silent, swimming nations of the scaly fish,
All the species of wild things, and all the happy herds
Of domesticated livestock, and the parti-coloured birds
That throng the fertile spots along the shore of lake and spring
And stream, and flock the pathless forests with their fluttering.
Then choose any individual and you will find
Each is unique, and in some way is different from its kind,
How else could any offspring know its dam, were this not true,
350 Or how could mothers know their young, which they so clearly
    do?
Animals, like people, recognize each other too.

For often before the graven shrines of gods, beside an altar
Smouldering with incense, a calf is offered for the slaughter,
And as he pants a steaming stream of blood out of his breast
His mother roams bereaved across green meadows without rest,
Looking for familiar cloven hoof prints in the ground
To see if there is anywhere her lost child can be found,

And standing in a leafy grove, she fills it with her low,
And keeps returning to their stall, stabbed with a yearning woe.   360
Not tender willow shoots, nor dewy grass, nor streams that
    wind,
Brimming to their banks, give any pleasure to her mind,
Or turn her stricken heart from grief, and neither can the sight
Of other calves in happy pastures make her burden light;
To such lengths does she seek after her own, and not another's.
Likewise, little kids go stuttering after their horned mothers,
And butting lambkins recognize the bleating of their ilk,
So each at Nature's call trots to the right udder of milk.   370

Lastly, look at any crop, and you will realize
Kernels of grain differ somewhat in their shape and size
Just like the quilt of seashells colouring the lap of land
Where the soft surf strikes against the shoreline's thirsty sand.
And seeing atoms are not manufactured from one mould,
But are a product of Nature, I declare as I have told
Before – it's a conclusion from which there is no escape –
The atoms of things that flit about must come in many a shape.   380

It's simple enough by reasoning of the mind for us to show
Why the fire of lightning penetrates more with its flow
Than fire that rises from the torches here on earth below:
For you could say of lightning's heavenly fire that it is more
Delicate and made of finer particles; therefore,
It passes through the openings that our *earthly* fire – brought
To birth and fashioned out of torches made of wood – cannot.
And then consider a lantern made of horn: it lets the light
Pass through, while it repels the water on a stormy night.
And how can this be possible unless it's also plain
That light is of a finer substance than the stuff of rain?   390
And we observe that wine will pour as fast as anything
Right through a colander, while olive oil will ooze and cling,
Either because the particles of olive oil are larger,
Or else because they are more hooked and knotted up together,
And so the particles cannot come suddenly undone
So that they can go dripping through the openings one by one.

And then consider milk and honey, liquids sweet to savour,
Rolled around the mouth, while there is such an evil flavour
400 To nasty wormwood and foul knapweed, on the other hand,
They pucker up the face. And so it's easy to understand
What's pleasant to the senses is of atoms smooth and round,
And what seems harsh or bitter is made up of atoms bound
Together by sharper hooks, so that these tend to rip and thresh
Their way into our senses and to tear into our flesh.

Lastly, what pleases or disgusts the senses with their touch,
These substances will clash with one another, seeing such
Differ in their shapes. For I should hope you don't divine
410 That elements that go to make a saw's hair-raising whine
Are smooth as elements that make a melody that sings
Wakened by the harpist's nimble fingers from the strings;
Nor should you think the particles that penetrate the nose
When we burn carcasses that have begun to decompose
Are the same shape as when saffron's just sprinkled on the
        stage[8]
Or nearby altar smokes with Arabian incense. Do not gauge
The elements that make the colours up that feast the eye
420 The same as those which prick the eye to tears, and make
        us cry,
Or which strike us as garish and ugly. For anything that can
        soothe
The senses must be made of atoms that are somewhat smooth;
While, on the other hand, the substances that irritate
Consist of rather rough material that tends to grate.
And there are particles which are neither, or at any rate
Not altogether smooth nor all a-bristle with sharp prickles,
Rather, with small jutting angles, the kind of thing that tickles
Rather than rips into the senses, for instance take the taste
430 Of dried wine lees or scabwort.[9] Burning fire and chilling frost
Both bite the body's senses with a different kind of 'fang',
As shown by how each makes us feel a different kind of pang.

For *touch* – yes, by the power of the gods! – *touch* is the key
To all our physical sensations, whether a feeling be

Triggered by a foreign substance worming its way within,
Or something born inside the body itself inflict a pain;
Or when, in the conceiving act of Love, the outward flow
Delights; or when the body's particles, struck by a blow,
Are shaken up, and throw the senses in confusion. No?
Then try this small experiment, and strike, with your own
    hand,   440
Any area of your body – then you'll understand.
Thus atoms come in many different shapes – we know it's true
Because the feelings they elicit differ widely too.

Now to resume: those substances which seem to be hard or
    dense
Must be made of matter held by close entanglements,
Interlacing branches, hooked together tight and snug.
Foremost among these is the diamond, which is wont to shrug
Off any shocks, and hard flint, and unbending iron ore,
And brazen hinges that protest the budging of a door
With piercing whines. But smoother, rounder atoms must
    compose   450
Any substance that is liquid, and whatever flows.
Consider poppy seeds – they pour as easily as water
Because the little spheres don't hitch or hinder one another,
And spill as easily downhill. And lastly, when you see
Something – smoke or mist or flame – that scatters instantly,
It must be made, if not of atoms utterly smooth and round,
At least not of *hooked* particles all tangled up and bound;
And since the elements of smoke can penetrate and sting
The senses, and those of fire can pierce stone, they must not
    cling   460
Together. Thus we easily see what 'bites' into the sense
Is made up, not of *tangled*, but, instead, *sharp* elements.

But that a single substance can have both fluidity
And bitterness at once – as with the salt-sweat of the sea –
Should come as no surprise. For even though salt-water's
    bound,
Being liquid, to consist of atoms smooth and round,

And mingled with rough particles that cause pain, yet it does
Not follow that these must be hooked and tangled up, because
Although rough on the surface, they are spherical in shape,
470 And so can roll across the senses even while they scrape.
It isn't difficult to show that smooth and rough combine
To make up Neptune's bitter body – that is, ocean brine –
For you can separate them, and see how the water, when
Filtered many times through earth, turns back to fresh again,
Flowing into a trap, the noxious brine thus put away,
Because the brine's rough particles cling tighter to the clay.

Now that I've demonstrated this,[10] I'll go on to explain
A fact whose proof is linked to it, connected in a chain:
480 *The number of the different shapes of atoms has a limit,*
Or else, again, the size of some would also be infinite.
For take the tiny confines of one atom: no very wide
Variety of shapes and forms is able to fit inside.
Let's say these primal particles themselves consist of three
Smallest parts – or make it five or six – and you will see,
When you arrange the atom's parts in various situations,
Change top and bottom, left and right, make all the
          permutations,
490 Then you'll discover every possible form that it could take.
And you will need yet more components if you wish to make
Other shapes than these; and further variety likewise
Would call for more components still. And so we realize
Novelty of forms is linked to greater atomic size.
And that is why it is impossible you should believe
Atoms have infinite shapes, that is, unless you also perceive
That some of them are immeasurably large in size, although
I have already demonstrated this cannot be so.

500 And now consider oriental robes and garments dyed
Glowing purple from the murex of the Thessalian tide,[11]
And the gilded race of peacocks dipped in iridescent grace –
All these would be eclipsed if some new colour took their
          place.

We'd snub myrrh's perfume, honey's flavour; the song the swan
    sings[12]
And Apollo's complex melodies plucked from the lyre's strings
Alike would be defeated and fall mute, because new things,
Each more wonderful than the last, would constantly arise.
Things might change for the better, but they might be
    otherwise,
And go from bad to worse, so every new thing that arose
Would disgust the senses more – offending ears and eyes and
    nose   510
And taste. But since this isn't so, and since the world is found
To have at either extreme a limit by which it is bound,
You must admit atomic shapes are finite in their number.

Another example: there is a limited range of temperature
Between the heat of fire at one end down to hoar-frost's chill –
Or measured from the opposite direction, if you will –
Everything from hot to cold, all luke-warms in between
Lie within these boundaries, and that is how they're seen
To make up the entire scale among them gradually.
Temperatures only vary by a limited degree
Coming between these two extremes – for they are bounded
    twice: 520
By flame at one end; at the other, paralysis of ice.

Now that I've demonstrated this, I'll go on to explain
A fact whose proof is linked to it, connected in a chain:
Atoms *similar* in shape have *no end* to their numbers.
Indeed, since there's a limited amount of different figures,
Those similar in shape must needs be numberless, or else
The sum of matter would be finite, which I've proven false
While showing in my poem that the universe abides
Because of the supply of atoms raining from all sides
In endless tandem blows out of the void.

                            For though you see   530
That certain animals, due to their low fertility,

Are relatively scarce, yet in another clime and place,
Some far-off land, there may be many numbers of their race.
Among four-legged beasts, this is especially the case
With snaky-handed elephants, whose myriads have made
A fence of tusks round India that no one can invade,
So massive is the population of those creatures *there*,
540 While seldom we set eyes upon them here, they are so rare.[13]

Or even say some one-of-a-kind being came to birth
So there was not another like it over the whole earth –
How could it be conceived and grow, and how could it arise
If matter did not come from inexhaustible supplies?
Or even were I to assume the particles it needs
Are taken from a finite pool of generative seeds
Tossing about the void, then how and where and by what
    motion
And force could they ever come together out in that vast ocean
550 Of matter, amidst such a riot of other atoms? Can they? No,
It is my opinion that this never could be so.
But as with major shipwrecks, when the vessel's torn to tatters,
And the mighty sea, as is her wont, tosses about and scatters
Floating transoms, ship ribs, sail-yards, spars, prow, masts,
    planks, oars,[14]
And bobbing pieces of the stern are littered on the shores
So that they seem to issue there a warning of the sea
To mortal men – her power and her snares and treachery –
To steer clear and beware of her, yes, even when she smiles
Enticing with serenity, duplicitous with guiles –
560 So once you have decided they are finite in their number,
Atoms in the same way must be tossed and torn asunder,
Pitching farther and farther apart on matter's seething tide
So they can never come into close contact and collide
To combine together into unions, nor can such abide,
Nor increase by adding on. Yet both are clearly so:
Things are indeed brought forth, and once they are brought
    forth, can grow.
Therefore it is clear there is an infinite supply
Of each kind of atom which all things can flourish by.

Neither can creation be forever overcome
By deadly blows, nor can those blows eternally entomb   570
Creation's vigour. Nor can the motions that combine together
To generate and make things grow, preserve them safe forever.
And so the war is ever being waged and ever tossed
Between creation and destruction, neither won nor lost –
Now here, now there, the forces of vitality prevail,
And now they are defeated in their turn. The funeral wail
Is mixed together with the sound of newborn babies' cries
When infants first behold the shores of light before their eyes,
And no night follows day, and no day from the night emerges,
Which does not hear the feeble mewls of babes mixed with the
      dirges
Which are the retinue of Death and of the dusky grave.   580

And here is one thing that it would be wise for you to save
Signed and sealed in the Bank of Memory: that you can find
Nothing that's composed of atoms of a single kind.
There's nothing that does not consist of different seeds
      combined,
And the more powers and qualities a thing has, so we find
The more it must contain shapes numerous and diverse. Behold:
The earth contains both particles which are the source of cold –
Hence gushing springs eternally refresh the boundless deep –   590
While also it contains the atoms from which fires leap;
For the ground, in many a place, is burning right beneath our
      feet,
While from the bowels of earth, Mount Aetna spews its raging
      heat.
The earth is also able to bring forth the shining sheaves
Of grain and fruitful trees for humankind, the streams and
      leaves
And fodder for the mountain-prowling beasts. Thus she is
      known
As the Great Mother,[15] Mother of Gods and Wild Beasts. She
      alone
Is creator of our bodies. And she is the goddess shown,

600 In songs of ancient Greek bards[16] learned in poetic science,
    Driving in a chariot drawn by a brace of lions.
    Thus poets symbolized that Earth rests not on earth but hangs
    Suspended in a space of air; they yoked wild beasts with fangs
    In order to represent that offspring, though they may be wild,
    Should be tamed by the kindness of their parents and made mild.
    And they portrayed her wearing fortifications for a crown,
    Since she is what upholds strategic ramparts of a town;
    These are the Holy Mother's attributes, and thus portrayed
    The world over, her effigy is carried in parade
610 With awesome pomp. The various nations in their ancient rites
    Hail her as the Idaean Mother and, for her acolytes,
    Give her a retinue of Phrygians, since people claim
    That it is first from Phrygia that agriculture came,
    And that from there it spread across the entire globe. They made
    Her priests eunuchs[17] to illustrate that those who would degrade
    The Mother's power, ungrateful to their parents, had no right
    To bring descendants forth unto the boundaries of Light.

    Tum, tum! – here comes the thrum of drums under open palms –
        and crash!
    The clash of cymbals all around her – bugles hoarse and brash
620 Bray their alarm, and hollow flutes stir frenzies as they play
    Phrygian rhythms. Heading the procession, a display
    Of flint blades represents her wrath in order to affright
    The crowd's ungrateful souls and impious hearts at such a sight
    With awe for her divinity. And as she's carried through
    Great cities, bestowing silent blessings unto men, they strew
    Her path with lavish riches, bronze and silver, and they snow
    Rose petals down to shade her and her escort as they go.
    And next, here comes a troop of youths in armour – it is these
630 Dancing among the Phrygians, whom the Greeks call 'Curetes' –
    Delirious with bloodshed, leaping to the beat, they quake
    The dread crests on their helmets when they give their heads a
        shake.
    These represent the Dictaean Curetes of ancient tales
    Who long ago in Crete disguised the baby Jupiter's wails,[18]

Dancing wildly round the infant, a circle of armed boys
Clashing bronze on bronze in rhythm, making a great noise
Lest Saturn get him in his clutches, gobble him up and deal
Such a wound to the Mother's heart that it could never heal.
And that's the reason why it is with weapons that the band   640
Escorts the Holy Mother, or because of her command –
Ready with arms and courage to defend the Fatherland –
So for their parents they became both guardian and glory.

A far cry from the truth, although it makes a pleasing story.
For godhead by its nature must enjoy eternal life
In utmost peace, removed from us and far from mortal strife,
Apart from any suffering, apart from any danger,
Powerful of itself, not needing us, and both a stranger   650
To our attempts to win it over and untouched by anger.
The earth is not a sentient being. The reason that the earth
Brings many different things in many different ways to birth
Into the sunlight is that within the earth there are supplies
Of many different elements from which these things arise.
If someone speaks of *Neptune* when he means the bounding
        main,
Or speaks of *Ceres* when what he's referring to is grain,
If he prefers to speak of wine by the name of 'Dionysus',
Instead of by its proper term, then it can be no crisis
If he calls earth the 'Mother of the Gods' – that is, as long
As he understands the superstitious element is wrong
And keeps his mind free from that taint.

                                        Often it comes to pass   660
That we behold, all grazing on the selfsame field of grass,
Flocks of woolly sheep along with horses bred for battle
Under the same tent of sky, and herds of hornèd cattle,
And all slaking their thirst at the same stream. Yet each remains
A separate and distinguishable species, and retains
The nature of its parents and behaviour of its race,
So various is the store of matter lurking in the grass
And such the diversity of matter in the stream. And then,
Furthermore, select from these a single specimen –

670 It's made of fluids, tissue, sinews, bones, blood, veins and heat –
Each of these a very different substance and discrete,
And thus composed of particles whose shapes are not the same.

Or consider anything that's kindled and devoured by flame:
If nothing else, at least it must have hidden in its members
Those atoms which give rise to fire, from which it glows with
        embers
And sends light shooting out and scatters sparkles everywhere.
Investigate other things and bring the same reason to bear:
Objects conceal the seeds of many things, as you will see,
And hold within themselves shapes of a wide variety.

680 Next, you'll notice many things have colour, smell and taste:
Consider all the sacrificial offerings [that are placed
On altars] – these consist of many different shapes, we know,
Since odour penetrates the flesh where colour does not go,
And colour enters the senses by one path, taste by another,
Showing the shapes of their components differ from each other.
Thus alien shapes convene into a single mass. Things need
To be composed of an amalgamation of mixed seed.

You'll notice scattered all throughout my verses as you read
Many letters common to an array of words – and yet
690 You must admit that different words, from this one alphabet,
Draw different letters. Which is not to say there aren't a few
Common letters running through the lot, or that no two
Words contain identical sets of letters, but we may claim
That as a general rule, all words cannot be spelled the same;
Thus many other objects may share elements in common,
And yet they may be held to differ in their composition.
And so it's right to say the race of mankind must be born
Composed of different elements from orchards or from corn.

700 Yet all things can't be joined every which way – for otherwise
You'd find that freaks were commonplace – a race of men would
        rise

Half-human and half-animal, or from their trunks would
    sprout
The limbs of trees instead of arms. Monsters would come about
With legs of terrestrial animals stuck to aquatic frames.
Chimaeras would arise, breathing poisonous fumes and flames,
Nourished by the all-engendering Earth. But since it's plain
Such prodigies don't happen, and because we ascertain
All things come from a certain dam and from specific seed,
And as they grow maintain the characteristics of their breed,
This happens by a fixed and certain process. For we find   710
Each thing retains the elements specific to its kind
From all the food it eats, and once these atoms are combined
They produce the necessary motions. But on the other hand,
The body naturally excretes and gives back to the land
Particles alien to it. Other elements as well
Transpire from the flesh in particles invisible
When they cannot connect with anything within or play
A part in vital motions and the blows drive them away.

But don't think only animals bound by these rules, because
Everything is kept in limits by these very laws.
For just as all things differ in their natures through and
    through,   720
They must be made of atoms that are shaped differently too.
Which isn't to say some substances might not share shapes the
    same,
But that in general they are different, it is fair to claim.
And since their shapes are different, it's reasonable to suppose
All else differs: intervals, connections, paths, weights, blows,
Meetings, motions – which not only keep the various species
Of animals distinct, but separate land from the seas,
And hold the firmament of heaven apart from the earth's soil.

Now pay attention to my words selected with sweet toil   730
Lest you should happen to surmise that what seems to your sight
To be the colour white is made of atoms that are white,
Or else assume that objects which seem black before your eyes
Have to be comprised of atoms that are black likewise –

In other words that objects of whatever shade you choose
Must be composed of atoms tinted in those very hues –
Because the atoms of matter possess *no colour whatsoever*,
Neither like the colours that we see, nor any other.
But if perchance you should believe the mind cannot apply
Itself to understanding what's invisible to the eye,
740 You stray far off the path. For even men who were born blind,
Who've never set eyes on the sun, nevertheless can find
Objects by touch, and recognize a thing that has been shorn
Of any link with colour from the day that they were born.
Thus bodies do not have to be imbued with any hue
For us to grasp a notion of them in our minds. It's true –
Consider how when we ourselves in pitch-black darkness clutch
An object, we can't tell the colour of it by its touch.

I've proved [things can exist deprived of hue; so it's no reach
To demonstrate atoms are colourless], as now I teach;
For every colour can be altered utterly into each.
750 But elements of things should not be changed in any way.
There must be something never alters – something has to stay
So all things do not utterly to nothingness decay.
For anything so changed it isn't bounded any more
By its own limits, dies at once to what it was before.
Think twice before you tint the atoms, lest all things regress,
And everything be utterly reduced to nothingness!

Besides, if atoms have no hue, but are endowed instead
With variations in their shapes and forms as I have said,
And they give rise to all the different colours of the rainbow
760 By changing their connections and positions where they go,
The motions they give or take, it's instantly easy to explain
How what seemed black just now, when we look at it once
    again,
Has suddenly turned white as marble; for instance, take the
    sight
Of the sea, when winds have churned the surface to a hoary
    white.
For you could say that often what appears as black as night,

When its matter has been mixed around and atoms rearranged –
Adding some, subtracting others – all at once is changed   770
To something gleaming white. But if the sea's water is made
Of deep-blue atoms, there's no way that they can change their
     shade
To white: however much you jumble objects that are blue,
Never into marble whiteness can they change their hue.
Or if you think that particles in a variety
Of colours make the pure and simple shining of the sea,
As sometimes we may notice that the figure of a square
Is made from other shapes – just keep in mind we are aware   780
Of the other shapes that make it up, and so we should likewise
Perceive the wide diversity of colours that comprise
The sea or other simply shining surfaces with our eyes.
Nothing about assorted different triangles would impair
Their union's ability to form the outline of a square,
And yet disparities of colour would prevent outright
The surface of a thing from shining uniformly bright.

The reasoning that sometimes lures us, leading us astray,
To endow atoms with colour collapses now in disarray,
Since white things are not made of atoms that are white, and
     too,   790
Black things are not from black, but atoms of assorted hue.
The truth is, whiteness will arise more easily from a lack
Of *any* colour than it will result from the colour black
Or any other hue that fights against and hinders white.

Moreover, since there can't be any colour without light,
And even in the light, atoms of things do not appear,
The atoms are not clothed in any colour, it is clear.
What colour can there be in blinding darkness? It's the play
Of light itself that changes colours, depending on the way
Its brightness bounces off them, at an angle or direct.   800
For instance, take the ring of plumage circling the neck
Of doves, how when it's struck by sun, the feathers seem to be,
Glimpsed from one angle, the red of fiery garnets, but we see
From another, green of emeralds mixed with lapis lazuli.

Or consider the peacock's tail flooded with sunlight – how a
    feather
When turned this way and that gives off one colour then
    another,
Each hue produced by an angle of the light. You should not
    doubt it –
And there is not a single hue could come about without it.

810 And since one kind of blow must strike the pupil to produce
The colour white, and another blow for black and other hues,
And since the only thing that matters to the touch is shape –
The colour makes no difference – then you cannot escape
The conclusion: atoms have no need of colour. Variations
In shapes of atoms bring about the different sensations.

Again, since shapes aren't limited to a specific hue,
And colours aren't assigned to certain shapes, it should be true
That atoms, if they did have colour, would come in every shade,
820 And likewise all the different things and creatures that they
    made.
For then it would be plausible to see a gleam of white
Flashing off the snowy plumes of ravens in mid-flight,
And there would be black swans composed of atoms dyed the
    same,
Or a whole rainbow of swans of any colour you could name![19]

Besides, the smaller the pieces into which you tear a thing,
The more you can perceive its colour slowly dwindling
Until it is snuffed out. For instance, if you rip to shreds
A wool cloth dyed Phoenician purple, as it's pulled to threads,
830 The purple, though the most intense of colours, starts to fail.
And in this way, you can discern that substances exhale
Their colour first before they are reduced to elements.

Lastly, since you must admit some substances lack scents
Or else are mute, then you cannot insist that sound or smell
Is granted to all bodies. And since you must admit as well
That there are substances which are to us invisible,

Then surely objects as bereft of colour can be found
As there are objects odourless and things divorced from sound:
The keen mind just as easily can grasp the likes of these   840
As notice things that are deprived of other qualities.
But colour isn't all the atoms lack. If truth be told,
They also have no temperature: not heat or warmth or cold.
They travel along sterile of sound, and starved of any flavour,
Nor do they send off from their bodies an inherent odour.
For just as in the preparation of a sweet perfume
From essences of marjoram and myrrh and spikenard bloom[20]
Which breathes a nectar to the nose, the first thing you must do
Is find as odourless an oil as you are able to   850
So that the olive oil as little as possible overpowers
With its own pungent smell the fragrances of myrrh and flowers
Mixed and concocted in it; likewise, atoms must not impose
An odour of their own upon those things which they compose,
For atoms cannot give off anything – cannot secrete
Sound or taste or temperature – not cold or warmth or heat –
All of which are properties that mortal things display –
Soft, porous or frangible, and subject to decay.   860
Atoms must give these properties wide berth if we would place
A permanent foundation under everything, a base
Holding up the universe, lest everything regress,
And you find all is utterly reduced to nothingness.

Next we must admit that animals that sense and feel
Are made of insensible particles. For this we can appeal
To examples that are lying manifest before our eyes,
Which make this obvious to us rather than otherwise,
They take us by the hand and show that animals arise   870
From things with *no sensation at all*. For instance, take the birth
Of living worms from filthy dung piles when the sodden earth
Festers with unseasonable rains. All things, moreover,
Transform themselves in the same way. And so rivers of water,
Leaves and fertile pastureland turn into herds of beasts,
And beasts become our flesh, our flesh in turn becomes the feasts
That strengthen savage brutes and raptors mighty-on-the-wing.
So Nature makes all nourishment into some living thing,

880 And fashions all the feelings of a creature from this food,
The same way she unfurls flames from dry tinder and turns
      wood
To fire. Now don't you see how much placements of atoms
      matter –
In what configurations they are bound up with each other,
What kind of motions they impart and take from one another?

And then, what goads the mind itself and makes the mind
      perceive
Conflicting feelings so that it forbids you to believe
That that which feels is born from what does not? You will
      assert,
No doubt, I should consider the case of sticks and stones and
      dirt –
890 Combined, they don't give rise to living sense, but stay inert.
But don't forget – I do not claim the elements that comprise
A sensate being must all necessarily give rise
In an instant to sensation. It matters a great deal what size
The bodies that compose a sensate being are, and then
Their shape, lastly, their movement, placement and
      configuration.
Such particles we can't discern when we see wood or earth,
And yet, when 'rotted' by the rain, these particles give birth
To tiny wormlings, since the ancient way they were arranged
900 Has been all stirred up by this revolution and is changed
In such a fashion it gives rise to life.

                                    But those who claim[21]
What feels is made from sensate elements, and that the same
Holds true for sensate elements, that these in turn derive
Their ability to feel from smaller things that are alive,
These people, by making atoms soft [must make them mortal
      too],
For all sensation's fastened to the flesh, to living tissue,
Sinews and veins, and since we note that these are soft, we see
They must be of a substance subject to mortality.

But even if, for the sake of argument, we were to say
That atoms such as these would last forever, not decay,
Either they must feel as separate parts, or must instead
Feel as entire animals. But parts, it must be said,   910
Cannot feel by themselves. It's evident that all sensations
In our body's limbs are bound in mutual relations.
If severed from us, neither the hand nor any part alone
Of the body is able to maintain sensation on its own.
So by default they must feel as whole animals, which makes
Atoms feel the same way that we feel, and each partakes
With us in that all-over feeling that is Life. But then
Who could say these are the building blocks of all creation,
Or that these can evade the pathways of annihilation
When they must be *living beings*. And 'alive' is just the same
As 'mortal' – the same quality called by a different name.
But even if, for the sake of argument, this were allowed,   920
A conglomerate of animals makes nothing but a crowd.
By just coming together, as should be obvious to you,
People, cattle and wild beasts cannot make something new.
Or if these atoms put aside their own sense in this case
And acquire another, why endow them with it in the first place?
Which brings us back to where I had left off – that since we
    catch
A clutch of bird's eggs turning into live chicks when they hatch
And worms squirming from soil torrential rains have made to
    rot,
We know that things that feel can come about from what does
    not.   930

But if someone by chance should claim that that which feels can
    rise
From what does not by metamorphosis, or otherwise
By some kind of birth through which sensation can appear,
These facts should be sufficient to demonstrate and to make
    clear
That no birth can take place without some prior combination,
And with no coming together, there can be no transformation.

For in the first place, there can be no feeling in the frame
Of any animal before formation of the same,
Because its elements are separate, and lie a-scatter
940 In air, streams, earth and products of the earth. And since this
        matter
Has not come together yet, it cannot get the right
Combinations of vital rhythms going to set alight
The watch-fires of the senses that keep in their oversight
All living beings.

                        Besides, take any creature – if a blow
Greater than it can withstand strike it, then this will throw
All senses, mind and body in disarray, and lay it low,
For it shakes the atoms loose from their positions, and will block
The vital movements utterly, till jolted by the shock
Throughout the flesh in every limb, the links of life unlock –
950 Those knots that bind the spirit to the flesh – so that the soul
Is cast out, scattered to the winds through every pore and hole.
For what else do we think that such a blow's able to do,
But strike a thing to pieces and dissolve it through and through?

Sometimes it happens that the blow is staggering but not fatal;
Then the surviving vital rhythms rally and prevail,
Prevail and settle the huge commotion of the blow, and make
Everything go back into its rightful path, and break
Death's motions, which already hold the body in their sway,
And rekindle the guttering fires of the senses. What other way
960 Can a creature come back from the threshold of Death's very
        door
To Life, with all its wits collected, instead of passing over
Beyond the finish line, the home-stretch almost run?

                                                    What's more
Since it is atoms shaken loose inside that trigger pain
When we receive a bodily shock, and their return again
Back to their rightful places that results in sweet delight,
You can be certain they can feel no pain in their own right,

And take no pleasure on their own, because they aren't
     composed
Of atoms by whose shifting motions they are indisposed,
Or elements whose changing movements let them pluck the
     fruit   970
Of pleasure's bounty. Therefore sense can't be an attribute
Of atoms! Again, if living beings can only be allowed
To have sensations if in turn their atoms are endowed
With feeling, what of atoms which compose the human race?
I suppose they shake with giggles, and that tears stream down
     their face
And dew their cheeks, and doubtless they are old hands at
     debate
On the make-up of the universe, and even investigate
What elements comprise themselves – for seeing they resemble
Entire human beings, they must be composed as well   980
Of smaller elements, and these composed of others still,
And on and on and on and on, *ad infinitum*, till
There's nowhere you can stop and make a stand – and even so
I shall pursue this argument however far you go –
For anything that you declare can speak, laugh or be wise,
Must be of atoms that do the same. But if we realize
This argument is raving lunacy and utterly daft,
And one can laugh and not be made of particles that laugh,
Be wise and skilled in philosophic discourse, while, however,
Not being made of atoms that are eloquent or clever,
Then why should not such things that feel result from
     combinations
Of particles that are themselves devoid of all sensations?   990

Finally, we all arise from seed celestial,
Because the same sky overhead is father of us all.
From him our nurturing mother, Earth, receives the rain's moist
     drops,
And gravid, she brings forth the joyful trees and gleaming crops,
And the human race as well, and every stripe of wild beast,
Since she provides the nourishment on which their bodies feast

And by which means they live sweet life, and bring their young
    to birth,
And therefore 'Mother' is a name that's fitting for the Earth.
For what arises from the earth falls back to earth once more,
1000 And that which was sent down to earth from heaven's aethereal
    shore
Is taken up again into the quarters of the sky.
Nor does Death demolish anything so utterly
That it annihilates the very atoms of its matter;
It only makes the combination of the atoms scatter.
From these it joins one particle with another – this is how
All things transform their shape and alter in their colour, now
Receiving sensation, and in an instant, yielding it up again.
That's how you know how much depends on the configuration
Of atoms, how they're held together, and in what position,
What motions they impart to one another or receive,
1010 And that's the reason you should not mistakenly believe
Atoms permanently retain those qualities we see
Sliding across the surfaces of things, both suddenly
Coming into being and of a sudden passing away.

Moreover, it is vital in what order I array
The different letters that make up my lines, in what position,
Because the *sky*, the *sea*, the *soil*, the *streams*, the shining *sun*,
Are drawn from a single pool of letters, and one alphabet
Spells *barley*, *bushes*, *beasts*, words not identical, and yet
With certain letters shared in common, for what really matters,
What makes a world of difference, is the *arrangement* of the
    letters.
The same goes for the physical. For when you rearrange
1021 Atoms, their order, shapes and motions, then you also change
What they compose.

             Now I need your full attention here –
A revolutionary thing strives hard to reach your ear,
A new side of the universe struggles to come to light –
For no fact is so simple we believe it at first sight,

And there is nothing that exists so great or marvellous
That over time mankind does not admire it less and less.
Behold the pure blue of the heavens, and all that they possess,   1030
The roving stars, the moon, the sun's light, brilliant and
     sublime –
Imagine if these were shown to men now for the first time,
Suddenly and with no warning. What could be declared
More wondrous than these miracles no one before had dared
Believe could even exist? Nothing. Nothing could be quite
As remarkable as this, so wonderful would be the sight.
Now, however, people hardly bother to lift their eyes
To the glittering heavens, they are so accustomed to the skies.
That's why you should let go of any terror of the new.   1040
But don't spit out my reason. Weigh with care. If it seems true,
What I'm about to say, then throw your hands up in surrender;
But if it should seem false, then arm yourself as truth's
     defender.
The mind seeks explanation. Since the universe extends
Forever out beyond those ramparts at which our world ends,
The mind forever yearns to peer into infinity,
To project beyond and outside of itself, and there soar free.

First, stretching all around us, limitless on every side,
Above, below, the universe extends out far and wide
Just as I've shown before, and as the fact itself cries out,   1050
And the nature of the fathomless Deep illuminates past doubt.
Since empty space is limitless on all sides and the amount
Of atoms meandering in the measureless universe, past count,
All flitting about in many different ways, endlessly hurled
In restless motion, it is *most unlikely* that this world,
This sky and rondure of the earth, was made *the only one*,
And all those atoms *out*side of our world get nothing done;
Especially since this world is the product of Nature, the
     happenstance
Of the seeds of things colliding into each other by pure chance
In every possible way, no aim in view, at random, blind,   1060
Till sooner or later certain atoms suddenly combined

So that they lay the warp to weave the cloth of mighty things:
Of earth, of sea, of sky, of all the species of living beings.
That's why I say you must admit that there are other cases
Of congregations of matter that exist in other places
Like this one here of ours the aether ardently embraces.

Besides, when matter is available in great supply,
Where there is space at hand, and nothing to be hindered by,
Things must happen and come to pass. That is a certainty.
1070 And if there are so many atoms now no one could count,
In all the time Life has existed for, the full amount,
If the same Force and the same Nature abide everywhere
To throw together atoms just as they're united *here*,
You must confess that there are other worlds with other races
Of people and other kinds of animals in other places.

Moreover, nothing in creation is the only one.
Nothing is born unique, to grow up, by itself, alone,
Without a species, but as one of a number of its kind.
1080 Put the example of the animals before your mind
And you will see it's so, with every mountain-prowling brute,
With Man's twin offspring, man and woman, with the scaled
        and mute
Nations of fish, with every feather of creature that can fly.
Therefore you must confess the same is true for earth and sky,
Sun and moon and sea and all the rest – none is the only
One of its kind, but there are countless others – they are rife,
For they too have a deep-set limit to their span of life;
They also have a body that's the product of a birth
Just as any teeming breed of creature here on earth.

1090 If you possess a firm grasp of these tenets, you will see
That Nature, rid of harsh taskmasters, all at once is free,
And everything she does, does on her own, so that gods play
No part. For by the holy hearts of gods, who while away
Their tranquil immortality in peace! – who can hold sway
Over the measureless universe? Who is there who can keep
Hold of the reins that curb the power of the fathomless deep?

Who can juggle all the heavens? And with celestial flame
Warm worlds to fruitfulness? And be all places at the same
Time for all eternity, to cast a shadow under
Dark banks of clouds, or quake a clear sky with the clap of
    thunder?   1100
What god would send down lightning to rend his own shrines
    asunder?
Or withdraw to rage in desert wastes, and there let those bolts
    fly
That often slay the innocent and pass the guilty by?

And since the start of time, the day the world was first begun,
The birthday of the sea and earth, the dawning of the sun,
Many atoms have been added on from every side,
Collected by the mighty universe's heaving tide,
So land and sea might grow from these, and the ceiling of the
    sky
Increase its loftiness and raise its rafters up on high,   1110
And air might tower overhead. The atoms that compose
Each substance are apportioned to it by the rain of blows
From every direction, being drawn to what's the same:[22]
Water adds to water, earth increases earth, and flame
Forges flame, air, air; till Nature, Maker of Things, brings each
To the zenith of its growth, the upper limit it can reach,
That stage when no more can be added to the vital veins
Than that which flows already in them, and now ebbs and
    drains.
And at this point, the lifespan of all things must reach its height,   1120
And Nature checks the reins of growth and pulls with all her
    might.
For everything that grows with giddy increase that you see
Scaling, by degrees, the ladder of maturity,
Assimilates more matter than it loses, just as long
As nourishment can be distributed with ease among
All its veins, and while there is not so much matter lost
Through its sprawling openings that it cannot make up the cost,
At its age, by feeding. We must concede, without a doubt,
That many particles are ebbing and are flowing out,

But atoms must be entering in greater numbers still,
1130 Until the object finally attains the pinnacle
Of growth. Then from that moment, little by little, over time,
Age breaks down the oak-strength and the vigour of its prime
And slides into decay. Indeed, once growth has reached its peak,
The larger and wider something is, the more that thing will leak,
Scattering bodies in all directions, nor can it sustain
The easy flow of nutrients it needs through every vein,
Nor is the supply of nutrients sufficient to restore
The waves of matter as they're gushing out. It is, therefore,
No wonder that things pass away, seeing how they become
1140 Enfeebled by the ebb of atoms, and since all succumb
To outside blows. At length, with age, the food supply will fail,
And lethal particles with their external blows assail
Ceaselessly, and pound things to submission and decay.
Indeed, the ramparts of the mighty world, in the same way,
Stormed from all sides, will tumble into crumbling ruin too.
For it is food that has to remake everything anew.
Food is what must prop things up, and food is what sustains
Everything. Yet it is a lost cause, because the veins
Cannot supply, and Nature is unable to provide,
The quantities of matter required for anything to abide.

1150 Even now, the world is past its prime, and the spent Earth
Can hardly fashion the scrawniest creatures, when she once gave
        birth
To all kinds, and the bodies of monstrous beasts.[23] I cannot hold
The race of mortal beings was lowered on a rope of gold[24]
To the fields down from the lofty heavens, nor that mortals
        came
From the sea, nor from the waves that smash the rocks. It's from
        the same
Earth that feeds them from her body now that they were born.[25]
Besides, she was the first who volunteered the gleaming corn
For Man, and joyful vines, sweet fruits and the green fields
        which now
1160 Can scarce be made to yield despite the salt-sweat of our brow.

We grind down oxen, the strength of farmers, the iron of the
    plough,
And still we barely eke a living from the fields – the soil,
So stingy with its fruits, has grown so greedy for our toil.

Now the old farmer shakes his head and groans again and again
That the hard labours of his hands have turned out all in vain.
He rails against the present, while he has nothing but praise
For the fortunes of his father – yes, those were the good old
    days!
And the planter of the spent and shrivelled vine-stock heaves a
    sigh,
Harps upon the times and shakes his fist against the sky.
He mutters how, in olden days, when men obeyed the gods,   1170
Folk could easily make a living from their narrow plots,
Seeing how, before, the farmers had much smaller lots.
He doesn't know that all things dwindle away, stage by stage.
All list towards the Rocks exhausted by the course of Age.

BOOK III

MORTALITY AND THE SOUL

You,[1] who first amidst such thick gloom could raise up so bright
A lantern, bringing everything that's good in life to light,
You I follow, Glory of the Greeks, and place my feet
Within your very footsteps. Not because I would compete
With you, but for the sake of love, because I long to follow
And long to emulate you. After all, why would a swallow
Strive with swans? How can a kid with legs that wobble catch
Up with the gallop of a horse? – the race would be no match.
As bees will sample every flower the blooming meadows hold,
10 So in your scriptures we devour all your words of gold,
Golden sayings, truly, that deserve to last forever.
For soon as your reason starts to preach about the Way Things
    Are –
To share these revelations that your god-like mind unfurled –
The mind's terror scatters, and the ramparts of this world
Fall away, so that throughout the whole of Space I see
The goings-on that come to pass. The gods appear to me
Enthroned in all their holiness and their serenity,
And where they dwell, wind never lashes them, cloud never
    rains,
20 And snowfall white and crisp with biting frost never profanes.
The canopy of aether over them is always bright
And unbeclouded, lavishing the laughter of its light.[2]
And there they want for nothing; every need, Nature supplies;
And nothing ever ruffles their peace of mind. Contrariwise,
The provinces of Hell are nowhere to be seen, although
The earth does not obstruct our view of everything below:
All in the void beneath our feet lies open to our sight.

Such revelations and I'm seized by a divine delight –
I shiver, for, due to your power, Nature everywhere
In every part lies open; all her secrets are laid bare.   30

And since I've taught what sort the rudiments of all things are,
And how, of their own accord they fly, in many a shape and
      size,
Driven into ceaseless motion, and how these can give rise
To all things in creation, the task that follows for me here
Is in my verses to explain and make the nature clear
Of mind and spirit, and toss that Dread of Death out on its ear,
Since that's what stirs the lives of mortals into such turmoil
From the very depths, and there is nothing that it does not soil
With the smirch of death, no pleasure, pure and clean, it does
      not spoil.   40
For when men say that illness or the loss of one's good name
Are more to be feared than death and the pit of Hell, and when
      they claim
To know the essence of the soul is blood, or even air[3]
If the fancy takes them, and our reason's something they can
      spare,
Look to their actions – you will realize these things they crow
Are not beliefs they base on proven facts, but just for show.

These same men having fled as exiles from their native lands,
Banished from the eyes of men, with blood upon their hands
From some dastardly deed, beset with every kind of woe,
Nevertheless keep living, and wherever these wretches go,   50
Appease their ancestors, and put black livestock[4] to the knife,
And send down offerings to spirits of the afterlife.
The more harsh that their hardships are, the harder they hold on
To superstition. To truly take the measure of a man,
You must observe him in the midst of trial and tribulation –
Then, from the bottom of their hearts, men say what they
      believe;
The mask is torn away, and what remains cannot deceive.

Take Avarice and the blind drive of Ambition:[5] both may draw
60 Wretched men to step outside the limits of the law –
Often even as partners and accomplices in crime –
As each man, day and night, strives harder than the next to
      climb
Atop the pyramid of power. It is largely the dread
Of death on which these open wounds of life thrive and are fed,
For Vile Disgrace and Bitter Want seem so far from the state
Of a sweet, established life, they almost loiter at Death's gate.[6]
Compelled by an unfounded fear, men, to evade such trouble,
70 Amass wealth by the blood of civil war, and they redouble
Their riches in their greed, heaping one murder on another.
Stone-hearted, they take pleasure in the sad death of a brother,
And are suspicious of the suppers served up by their kin.

Likewise, envy, sprung from the selfsame fear, worries them
      thin:
Why should *that man* win power? *That man there* before their
      eyes?
And be looked up to, strutting in the bright robes of high office –
They gripe – while *they* writhe in the mire of obscurity and
      shame?
Some fritter their lives away pursuing statues and a name.
Sometimes the phobia of death can grip a man so tight
80 He comes to loathe his very life and looking on the light,
And in his mournful heart resolves to die by his own hand,
Oblivious this fear's the source of what he cannot stand –
That *this* leads one to outrage honour, another to break the ties
Of friendship, and in short to overturn all loyalties.
For there are many instances where people have betrayed
Their fatherland and their beloved parents to evade
The realms of Hell. And just as children, trembling, are afraid
Of all things in the dark, so sometimes we in broad daylight
90 Fear things imaginary as what babes dread in the night.
Thus we must put the shadowy terrors of the mind to flight,
Not by the illumination of the sun and its bright rays,
But by observing Nature's laws and looking on her face.

First of all, I say that what is called the intellect
Or mind, the seat of reason and the rudder to direct
Our lives, is part and parcel of a man, no less than feet
Or hands and eyes, the parts that make a living thing complete.
[. . .
Some hold] the mind in no set region of the flesh resides,
But as some essential condition of the body it abides,
A state the Greeks call 'harmony',[7] and they claim it is this   100
Phenomenon that animates and gives us consciousness,
Although the mind in no particular organ can be found.
Just as we often hear a body called healthy and sound,
Yet 'healthiness' is not a mere part of a healthy person,
So in no single part do they pinpoint the mind's sensation.
Here is where they make a wrong turn, and their reason fails.
For often, right before our eyes, we see our body ails,
While on the other hand our inner life is right as rain.
Often enough it is the opposite: the mind in pain,
While the body's hale and happy. And the same thing can be
    said   110
Of a wretch with throbbing foot but with no aching in the head.
Furthermore, when limbs surrender unto tender sleep,
And the body sprawls unconscious, slumbering heavily and
    deep,
Even at that hour, there is something in us still,
Restive and agitated in many ways, susceptible
To all the stirs of joy and bootless sorrows of the heart.

In order that you understand that *spirit*, for a start,
Likewise is seated in the flesh, and the ability
To feel does not come from the body's parts in harmony:
Consider that a body can have members shorn away –
Yet even when it's badly maimed, the life will often stay.   120
But on the other hand, when a few particles of heat
Scatter abroad, and air leaves through the mouth, life will
    retreat,
And in a trice vacates the veins, abandoning the bones.
This shows you that all particles do not share the same
    functions,

Nor are they equally vital to sustaining life. But, rather,
Those that are the seeds of wind and seeds of heat take care
Life lingers in our flesh. Therefore there is a vital breath
And warmth within our body that deserts it right at death.

130   Thus since the nature of the mind and spirit, as we learn,
Is basically a *part* of a human being – then return
The name of 'harmony' – a word that should belong by rights
To the musicians, something fetched down for them from the
      heights
Of Helicon, or maybe they got it somewhere else, and came
To appropriate it for their art, which needed its own name –
Whatever the case, *let them have it.*

                                    Now pay attention here:
I tell you mind and spirit[8] are bound up with one another,
And that together they combine to form a single nature.
But what heads the whole body and reigns over it like a king
Is Judgement, which we also name the 'mind' or 'understanding',
140   And this keeps its abode in the mid-region of the chest.
Here is the place that fright and dread stampede, here also rest
Soothing joys, so this is the region where the judgement stays.
(The remainder of the spirit, seeded throughout the flesh, obeys,
And a mere nod or signal from the mind puts it in motion.)

The mind is thoughtful by itself, rejoices on its own
When nothing agitates the flesh and spirit. Just as pains
Afflict our head or eyes, while all the rest of us remains
Free from torment, so the mind alone may suffer hurt
150   Or bloom with joy, while the remaining spirit stays inert
Throughout the flesh and feels no new sensation. But if fear
More strongly jolts the mind, the spirit feels it too, it's clear,
All through the frame. Hence the entire body sweats and pales,
The tongue loses its strength, the voice breaks off and speaking
      fails,
The eyes go dark, the ears buzz and the limbs fold and give
      way.[9]
In short, men often physically collapse beneath the sway

Of mental dread. From this it's easy for anyone to discern
That spirit's yoked to mind – for struck by the mind's power, in
	turn,
The spirit strikes and drives the body forward. And we learn		160
By means of this same reasoning that the nature of the mind
And of the spirit is a physical one – for when we find
That they propel the limbs, and snatch the frame from sleep, and
	change
The expressions of the face, and steer and govern all the range
Of movement in a person, it is obvious that such
Actions can be brought about *only by means of touch*.
And since what 'touches' must be material, isn't it true
That mind and spirit must be physical in nature too?
Besides, the mind and flesh suffer together, it is clear,
And feel in sympathy – for take the quivering blow of a spear		170
That misses vital organs, but lays bare sinew and bone –
The wounded man's still overtaken by a giddy swoon;
Earthward he sinks, with all his wits at sea, while now and then
He has a vague desire to struggle to his feet again.
The nature of the mind is physical, as all this shows,
Since mind's afflicted by real weapons and by tangible blows.

Now I shall undertake to demonstrate to you what kind
Of material makes up the composition of the mind.
First, I assert the mind is delicate as it can be,
Of particles surpassingly minute. And you can see			180
It's true by turning your attention to the following facts.
Nothing is speedier than the mind. We see that nothing acts
As fast as the mind imagines and initiates. We find,
Therefore, that nothing rouses itself as quickly as the mind,
And what moves very easily must be composed of such
Rounded, very tiny particles, the slightest nudge
Sets it in motion. Take water, how it ripples at a touch
Since it consists of tiny shapes that tumble easily.			190
And then compare the clingier viscosity of honey;
Its liquid is more sluggish and is more grudging to pour,
Because all its constituent matter sticks together more.

This is, without a doubt, because the bodies that are found
To make up honey, are not so smooth, so delicate, or round.
Or take a towering pyramid of poppy seeds – you know
The merest puff of breath can send the peak spilling below;
Contrariwise, a mound of stones or heap of ears of wheat
Won't budge.[10] And thus as bodies are very smooth and round
        and neat,
200 So they enjoy mobility; but on the other hand,
The heavier something is, and rougher, the firmer it will stand.
Now, since the mind moves with especial ease, as we have
        found,
It must be made of bodies exceedingly smooth and fine and
        round.
And once you understand this fact, my friend, then it can shed
Light on many things – and it will stand you in good stead.

Another proof that indicates the nature of the mind –
How fine its fabric is, and how the whole could be confined,
210 If it were gathered up, into a space of tiny span –
Is that, right when the peace of death has laid hold of a man
And mind and spirit retreat, inspect the body how you may,
You won't discover anything at all is taken away,
Neither in aspect nor in weight. For Death keeps the complete
Appearances of Life, except for vital sense and heat.
And therefore the whole soul[11] consists of very tiny seeds
Bound up in sinews, flesh and veins; for once the soul recedes
Entirely from the body, still the flesh does not renounce
220 The intact outline of its limbs, nor does it lose an ounce.
And it is much the same when vintages lose their bouquet
Or when sweet breath of perfumed oil evaporates away
Into the air, or when a substance loses all its smack
And yet our eyes discern no diminution and no lack
And nothing is subtracted from its weight – this clearly leads
To the fact that taste and smell are made of many tiny seeds
Scattered throughout the substance. Thus I once again insist
The nature of the mind and spirit is such it must consist
Of stuff composed of seeds that are so negligibly small,
230 Subtracted from the flesh, they don't affect the weight at all.

Nor should we think this substance is composed of one thing,
    neat,
For from the dying there escapes a slight *breath* mixed with *heat*,
While heat, in turn, must carry *air* along with it; for there
Is never any heat that is not also mixed with air,
Because heat's substance, being loose in texture, has to leave
Space for many seeds of air to travel through its weave.
This demonstrates the nature of the mind's at least threefold[12] –
Even so, these three together aren't enough, all told,
To generate sensation, since the mind rejects the notion
Any of these is able to produce sense-giving motion,
Or the thoughts the mind itself turns over. And so to these same   240
Three elements, we have to add a fourth that has no name.
There is nothing nimbler than this element at all –
Nothing is as fine as this is, or as smooth or small.
It's this that first distributes motions through the frame that lead
To sense, since this is first to bestir, composed of minute seed.
Next heat takes on the motion, then the wind's invisible force,
Then, the air, then all is set in motion in due course –
The blood throbs, flesh is all a-tingle, and at last a measure
Of feeling reaches down to the bones' marrow – either pleasure,   250
Or else the opposite, arousal. Pain can't easily seep
So far within, nor bitter injury penetrate so deep,
Without tending to throw the entire frame into such rife
Confusion that there isn't any place left there for life,
And the particles of spirit scatter abroad through every pore.
But such commotion rarely breaches the body's surface, for
We could not otherwise stay hale and hearty and alive.

How these four are mingled and arranged so that they thrive
I ardently wish to clarify for you – it is just the lack
Of terminology in our mother tongue that holds me back.   260
But as far as I am able, I'll outline the arguments:
The atoms that make up each of the spirit's elements
So mingle in their motions, no one element alone
Can be isolated, nor can an element act when on its own;
They are the many aspects of a single entity.
Just as, with the body of any living creature there will be

A scent, a certain warmth and flavour, and yet these enmesh
Together to compose amongst themselves a single flesh,
So heat and air and the unseen power of wind together mingle,
270 Along with that other quick and nimble power, to make a single
Essence, and it is this other power that imparts
To the others the impetus of motion whence sensation starts
And spreads throughout the flesh. For this fourth nature lurks
    and hides
Deep inside us; there is nothing deeper that abides
Within our bodies. It is the very soul of all the soul.
Just as, mingled in our limbs and bodies as a whole,
The force of mind and power of spirit lie hid from our view,
Being made of particles so tiny and so few;
Likewise, composed of tiny bodies, this power with no name
280 Lies hidden, a sort of spirit of the spirit, and this same
Power furthermore holds its dominion over the frame.
Similarly, the forces of air and wind and heat derive
Their power from their combination; tangled up they thrive
Throughout the limbs, one yielding while another's on the rise,
So together they all form a unity, for otherwise –
If heat and wind were separate, and sundered from air's force –
They would disperse sensation and destroy it by divorce.

The mind too has a measure of heat – as when it seethes with
    ire,
And from the eyes you see the flashes of a fiercer fire.
290 It also has its share of chilly wind, fear's partner – take
The way it gives you gooseflesh and it makes the body shake.
As well, the mind possesses the tranquillity of air,
When the heart's at peace, for instance, and the countenance is
    fair.
But in the nature of some animals, heat tips the scales –
In their fierce minds and savage hearts, wrath boils up and
    prevails.
The violence of lions springs to mind more than the rest;
They constantly have snarls and roars erupting from their
    breast,

Because their spate of rage is such it cannot be confined;
While a preponderance of wind makes up the chilly mind
Of deer, sending its icy gusts more swiftly through the frame,   300
And wind's what sets their limbs to trembling. It's not the same
With cattle: placid air's the vital portion of their nature –
Neither easily kindled by the smoking torch of anger
That casts its pitchy cloud, nor frozen by cold darts of fear.
They hold the middle ground, and stand between fierce lions and
    deer.

So too with men. Though education give them equal polish,
Still there are traces of their nature nothing can abolish.
Character flaws can't be uprooted – there will always be   310
One fellow who is more inclined to fall into a fury,
Another who by terror is more rapidly unmanned,
A third who passively puts up with more than he should stand.
The characters of men differ in many other respects
As well as those behaviours that the character affects –
But now I cannot list the secret causes that are to blame,
Nor invent a slew of terms that's large enough to name
The divers shapes of atoms whence such variation came.
But one thing I am certain of, so weak is any trace
Of inborn nature past the power of reason to erase,   320
That there is nothing that is fundamentally at odds
With living out our lives so they are worthy of the gods.

This nature, then, is found throughout the body as a whole.
It is the body's sentinel, what keeps it sound. For soul
And body cling together with common roots, and so we see
The two cannot be wrenched apart without calamity.
Just as it would be difficult to pry away the smell
From nuggets of frankincense without destroying them as well,
So it is hard to extricate the spirit and the mind   330
From flesh without undoing everything that's left behind –
So tangled are their elements from the first hour that begot
The life they live together and that is their common lot.
We see that neither body nor the mind contains alone,
Divorced from one another, the power to feel all on its own.

But together they produce communal motions, lighting the
    flame
Of feeling for us, fanning the senses' fires through the frame.

Furthermore, the body does not by itself arise,
Nor can it grow all on its own, nor last after it dies.
For it is not like water, which often gives off added heat,
340 Yet is not torn apart by this, but stays intact, complete.
Not so the body. Once the soul and flesh have parted ways,
The body left behind unravels utterly and decays.
The interlacings of the soul and body start to learn
Life's essential motions even before they have been born –
From the moment of conception, while still in the mother's
    womb.
The two cannot be sundered without leading to their doom.
And since they must be joined together in order to survive,
Their nature too must be conjoined, and that is how they thrive.

350 If anyone denies the flesh feels, but instead believes
That spirit mingled through the entire frame is what receives
The motion that we call sensation, he takes up a fight
Against facts that are clearly true and obviously right.
For who can elucidate sensation – what it is to feel –
Except the senses themselves, the things they teach us and reveal.
'But when the spirit leaves the flesh, sensation too is gone.'
True, and yet flesh loses what in life was not its own,
Like many other qualities[13] it loses when it dies.

Furthermore, to claim that we see nothing with the eyes,
360 But that the mind peers out through them as out of opened
    doors[14]
Is problematic, when their very feeling is the source
Of the opposite conclusion. Their own feeling tugs us right
To the eyes, especially since we can't look on what is bright
Because the dazzling light obstructs the clearness of our sight –
Clearly *not* the case with doorways – for an open door,
One we're looking out of, never squints because it's sore.

Indeed, if eyeballs serve as open doors, why then, no doubt
The mind would have a better view of things with them plucked
        out –
Tearing down the doors, and even the doorposts in the way!

Another error into which you never ought to stray,   370
Though set forth in the scripture of Democritus, the great
Philosopher: that atoms of flesh and spirit alternate,
One then the other, thus knitting the fabric of the flesh. For since
The elements of soul are smaller than the elements
Of flesh in size, so also they are fewer in their numbers,
And are distributed but sparsely through the body's members.

Here is something you can be assured of, for a start:
The space between soul-particles is *at least* as far apart
As the size of the slightest particles that stimulate sensation.   380
For at times we do not feel a film of dust stick to the skin,
Nor, lighting on our limbs, the sprinkling of powdered chalk,
Nor feel the fog at night; sometimes we're netted as we walk
Right through a spider's web, and do not feel the flimsy threads,
Nor feel its shrivelled remnant dropping down upon our heads,
Nor the feathers of birds, nor the down of thistles blowing
(Things so light they find the act of falling heavy going),
Nor every creepy-crawly thing that inches by degree,
Nor pitter-patter of each footstep of a gnat or flea.   390
And that's because a lot of atoms must be moved inside
Our frame before the atoms of the spirit, scattered wide
Throughout the flesh, can start to feel the motion, and before
They leap the great divide and meet and bounce apart once
        more.

It's more the mind that keeps life safely under lock and key;
The mind more than the spirit has life in its sovereignty.
For when the mind has fled, no shred of spirit can remain
In the flesh a moment longer, but follows swept up in its train,   400
Dispersing into air, and leaving the icy limbs behind
In the frosty grip of Death. He stays alive in whom the mind

Remains; for take a man whose limbs are hacked off, and who's
    left
As just a mangled torso, and likewise has been bereft
Of spirit with his amputated limbs, yet even so
He *lives* and breathes the life-affirming air, and even though
Deprived of most, if not all, of his spirit, clings to life –
Just as, if someone cuts around an eyeball with a knife,
But does not touch the pupil, the power of sight survives.
    (Unless
410 He happens to demolish the whole eyeball in the process
By cutting out the pupil, which will cause both parts to die.)
If just that little dot there in the middle of the eye
Is eaten away, the eye's light sets at once, the shadows fall,
Though otherwise there is no damage to the glittering ball.
Such is the pact that binds spirit and mind once and for all.

Now then, listen. In order for you to fully comprehend
That minds and flimsy spirits have a birthday and an end,
I've spent long hours hunting the right words, a labour of love,
420 To set this forth for you in poetry that's worthy of
Your life's calling. But do this favour for me just the same,
And yoke both of these concepts underneath a single name,
So that, say, when I speak of *spirit*, teaching that it dies,
Understand I am referring to the *mind* likewise,
Seeing that a single soul is formed out of their union.

First, because I've shown its texture is a gauzy one,
And that its particles are tiny – tinier for that matter
Than are the particles of fog, or smoke, or liquid water –
For it is nimbler by far, moves at the slightest jog,
430 Being triggered by mere *images* of smoke and fog,
As, for instance, when we're deep in sleep, and in a dream
We look on altars breathing out their smoke and sending steam
Aloft. (These images are carried to us, there's no doubt.)
And since you know how when a vase is shattered it pours out
All of the liquid it was holding, spilling it everywhere,
And since you know how fog and smoke disperse into the air –

Trust that the spirit pours and scatters at even greater speed,
And it resolves more quickly into its component seed
At the instant it departs and flees the body. On the whole,
If the body, which is, as it were, the vessel of the soul,   440
Made rarefied and leaky by the gouts of blood that drain
From veins when it is cracked somehow, no longer can contain
The spirit, how can *air* contain it, when air is a mesh
More full of holes than any leaky vessel of the flesh?

Body and mind are born together – we sense that this is true –
And they mature together, and grow old together too.
For just as children's frames are weak and wobble, being new,
So too their intellect is soft, and toddling along;
And when they're of an age where they have grown robust and
    strong,
Their judgement has grown also, and the power of the mind.   450
Later, when the body's worn down by Time's crushing grind,
The frame stoops and the limbs go feeble, the intellect goes
    lame,
The tongue gibbers, the mind totters. At once, all fails the same.
So it is only natural this spirit of which I spoke
Disperses, scattered on the high winds like a puff of smoke.
Seeing that together flesh and spirit are begot,
And that they both grow up together, just as I have taught,
And Old Age wears them down alike and fritters them away.

In addition to all this, as we see that the flesh falls prey
To horrible disease and pain that's difficult to bear,
So too the mind is prone to sorrow, terror, bitter care.   460
It follows, therefore, that the mind must also have a share
In Death. Indeed, the mind, when in the throes of physical pain,
Often wanders astray and raves, delirious, insane.
At other times, the mind is carried off into a deep
Coma, and sinks down into a never-ending sleep:
The eyes rolled back, the head nodding, it cannot hear the
    sound
Of voices, cannot recognize the people gathered round

Pleading it to return to life, their faces wet with tears.
470 And so it's plain the mind decays as well, since it appears
Disease can enter the mind with its contagion and its strain.
For these two are the architects of Death: Disease and Pain;
As we well know, taught by the deaths of many who've gone
        before.

Another thing: when wine has soaked a fellow to the core,
And through his veins has scattered and distributed its heat,
Why do his limbs grow leaden, why does he trip on his own feet?
Why does his mind sop, his sight swim, his tongue drawl?
480 Why does he burst with bellowing, and belching and a brawl,
And all the usual things that follow on a drunken spree?[15]
Why indeed, unless because the wine's ferocity
Tends to disrupt the spirit even while it's in the frame.
It's clear that if a thing can be disrupted, then the same,
If penetrated by a force of slightly stronger sway,
Will be deprived of any future and will pass away.

Often, before our eyes, a person suddenly is hit,
As with a bolt of lightning, by an epileptic fit:
He collapses foaming at the mouth, he moans, his body shakes,
490 He babbles and his sinews strain. He twists and turns. He takes
Ragged gasps, and by his writhing, wears his muscles out.
This is because the force of the disease, without a doubt,
Scrambled throughout the flesh, whips up the spirit so it
        breathes
In heaving gasps, and foams, just as the salty ocean seethes
When the waves are set to roiling by the blasts of powerful gales.
The moans are squeezed out when, throughout the limbs, a pain
        assails
The flesh, and also because the seeds of voice will tend to all
Come rushing out of the mouth, rolled up together in a ball,
Following the path their passing has made smooth before.
The raving comes from mind and spirit thrown into an uproar,
500 And from their being pulled asunder, rent and split in twain,
As I have demonstrated earlier, by that selfsame bane.
Next, when the cause of the disease already's on the wane,

And the biting humour of the ailing flesh is in retreat,
Recoiling to its lairs, the person struggles to his feet,
Staggering, returning to his senses by degrees,
Getting his spirit back again.[16] Since such a harsh disease
Shakes up the mind and spirit so, and pulls apart the same
To pitiful pieces while they're still inside the body's frame,
Then how can you believe that they are able to abide
Without the body, at the mercy of high winds outside?

And since we realize that medicine affects and heals   510
The mind as well as ailing flesh, this evidence reveals
The living mind is mortal, since whoever wants to change
The mind or any other substance has to rearrange
The organization of its structure, or add to the sum,
Or else must take away at least some tiny morsel from
The whole. But what's immortal does not suffer any new
Arrangement of its members, nor can it be added to,
Neither can even one iota of it flow away.
For anything that does, because of transformation, stray
Beyond the limits of itself, then from that moment on
Whatever thing it might have been *before* is dead and gone. 520
Therefore, if the mind can sicken, or if it can be
Affected by a drug, that's proof of its mortality,
As I have demonstrated before, and so Truth runs to meet
False Reasoning, and battles it, and cuts off all retreat,
And Falsehood, on the two-edged argument, meets its defeat.

Besides, we often see that it is piecemeal men succumb,
One limb after another losing life and falling numb:
First of all, the toenails and the toes go blue and black,
The feet and legs are next to die, then chill death leaves its track
As it creeps across the other members. And thus because the
      spirit   530
Is divided up and does not, when it leaves the body, clear it
All in one piece, then it is mortal too. If you should think
The spirit has the ability to retract itself and shrink
Into a single spot and pull its particles together
And so withdraw sensation from one limb after another,

Consider that the place in which the spirit then condenses
Should have, by rights, a corresponding heightening of the
    senses;
But seeing that there's *no such place*, again I must declare,
It perishes, being torn to shreds and scattered to the air.
540 And even if, just for the sake of argument, I grant
That spirit *can* be concentrated (though in truth it can't)
In the flesh of those who leave the Light by dying bit by bit –
The spirit's mortality is something you must *still* admit.
For whether the spirit perishes abroad, for winds to scatter,
Or shrinks up in a ball and goes inert, it does not matter –
Either way, sensation fails the man on every side,
And everywhere there's less and less life in him to abide.

And since the mind's a part of man, and has its fixed location,
550 As eyes or ears or other organs for life's navigation,
And just as hand, or eye, or nose, if separate and free
From us, would neither have the power to feel nor even *be*
But, rather, in short order would be melted with decay,
So mind requires the body – the actual man – in the same way
In order to exist, because the flesh contains the mind –
The body being, as it were, a vessel of a kind –
Or maybe there's some other metaphor that makes it plainer,
Since mind and flesh are closer bound than contents and
    container.

To resume, the lively power of mind and body only thrive
In partnership with one another – it's how they stay alive.
560 The mind without the body is unable to produce
Life-motions; drained of spirit, the body can't endure, or use
The senses. As the eye can have no power of sight plucked out
By the roots and unattached to any body – there's no doubt –
So we see mind and spirit have no power on their own,
Clearly because they're mixed with flesh and vein, sinew and
    bone,
And contained throughout the body as a whole. While shut
    inside
A narrow space together, they can't easily leap wide

Apart from one another, and so are able to produce
Life-motions, which they could not do outside the body, loose,   570
Scattered after death to the four winds, since they no more
Would be in the close quarters where they were contained
    before.

For the air itself would be a body – one that was alive –
If the spirit could hold itself together there, and could contrive
To make the motions that it used to make in flesh and bone.
I say again that once the body's covering comes undone
And the life-breath is expelled abroad, then, you must grant it's
    true,
The consciousness of mind unravels, and the spirit too;
Body and soul only exist because they're yoked together.
And since the body can't exist if it is ripped asunder   580
From spirit without decomposing, reeking of decay,
It's clear that spirit, risen from its depths, has seeped away
And scattered like a puff of smoke, and that what is to blame
For the change and the dilapidation of the body's frame
Is that its deep foundations have been shaken to the core
When the spirit oozed out of the limbs and fled through every
    pore
And twisting passage of the flesh. By all these ways and more
You can discern the spirit's substance has already scattered
Asunder when it leaves the flesh, and that it has been tattered
And torn while still within the flesh before it slips outside   590
The body and it floats away on the air's windy tide.

Indeed, while still contained in life, the spirit, even so,
If shaken loose by something, often seems to want to go
And to be freed from the body altogether. Then the cast
Of the countenance goes slack, as if this hour is its last,
And limbs go limp and dangle from the bloodless frame. It's
    this
We mean when we say someone 'passes out' of consciousness,
When everybody standing round, and trembling with fright,
Wants to grab onto the last link of Life and hold on tight.

600 For then the mind and spirit's power are rocked through and
          through
    And on the brink of giving way, the body weakened too,
    So that a slightly stronger shock could level them. Why doubt,
    In that case, that the spirit expelled from the body, driven out
    Into the open, weak and naked, not only lacks the power
    To endure forever, but cannot last a sliver of an hour?

    Nor does a person seem to feel his spirit, as he dies,
    Leave the body in one piece, nor does he feel it rise
    First to the throat, then travel up and exit through the jaws,
610 But feels it fail right in the fixed location that it was,
    Just as he knows the other senses fail where they remain.
    But were the mind immortal, then it would not so complain
    As it passed away, of scattering; but, rather, it would take
    Its going as release, shedding its garment like a snake!

    Again, why are the mind and its intelligence never begot
    In head or feet or hands – why do they stick in the same spot
    In a fixed locale for all alike, unless it is the case
    That each thing is allotted, for its birth, a certain place,
    And once it's born is able to survive where it is found.
620 Each has its parts arranged just so, never the wrong way round,
    With such consistency does one thing from another rise:
    Flame is not made from rivers, nor does fire give birth to ice.

    And if the spirit's nature is immortal, and we claim
    That it can feel sensation even separate from our frame,
    I suppose we must endow it with the five senses as well –
    How else do we envision phantoms roving down in Hell? –
    And that's the reason painters and the bards of yesteryear
630 Invented shades supplied with all the senses. But it's clear
    A spirit can't possess an eye or hand or nose or ear
    Or tongue without a body. There's no possibility
    That spirits by themselves can feel; indeed, they cannot *be*.
    And since we feel vital sensation is located throughout
    The body, that the whole of it's alive, without a doubt

If suddenly some violence were to slash the body through
The middle with a rapid stroke and slice it right in two,
The spirit would be cut in half and sundered just same,
Cleft and chopped in pieces at the same time with the frame,
But what can be divided into parts, what you can sever,   640
Obviously relinquishes all claims to last forever.

Chariots tricked out with scythes and cutting a bloody swath
Of steaming carnage often lop a limb off in their path
So suddenly it quivers on the ground right where it fell,
Sundered from the body, while its owner can't yet tell
What's happened, from the swiftness of the shock, and feels no
      pain –
Also because the lust for battle overpowers his brain.
With what's left of his body, he seeks slaughter and the field,
Not grasping that his left arm has gone missing with its shield,
Snatched away in a whir of horses, wheels, bloodthirsty knives.   650
Another, unaware his right arm's fallen off, still strives
To climb back in the fray. Another tries to make a stand
On one leg, while the dying leg, beside him in the sand,
Twitches its toes. Even the head, though lying in the dirt,
Hacked off from the hot and living trunk, keeps its alert
Countenance and open eyes, the look of life almost,
Until it has surrendered up the last scrap of the ghost.

Or else take the example of the serpent if you like –
Its flicking tongue, its threatening tail, its length reared up to
      strike –
If you chop up its body into pieces with your blade,
You'll see the severed segments writhe while wounds are newly
      made   660
And splatter the dirt with gore, and you will see the front end
      turn
Back to try to bite itself, smarting from the burn
Of the wound. But if we say that every piece of snake contains
An entire spirit, it follows that one animal retains
Many spirits in its body. Instead it is the same
Single spirit that is hacked to pieces with the frame.

Flesh and spirit both are mortal – that much is decided,
Since both can be cut into many pieces and divided.

670 And if the nature of the soul's immortal, and it creeps
Into the body as we're born, why is it no one keeps
A memory of time before, why can't we bring to mind
The deeds that we have done, why do they leave no trace
        behind?
For if the mind has undergone a transformation vast
Enough to cancel out all the remembrance of things past,
That is a state approaching little short of death, I'd say.
Therefore you must admit that what it *was* has passed away,
And that which it is *presently* has been created fresh.

And if the mind's life-force is added when the body's flesh
680 Is already complete, when we are born and cross the thresh-
Hold of Life, why then the mind should not appear to grow
Together with the body, limbs, and even the blood. No –
Instead, it should live by itself, alone, as in a cell
(While nevertheless sensations could flood through the flesh as
        well).
Thus I repeat, you must admit that spirits cannot be
Without beginnings, nor are they immune from Death's Decree.
For we must not believe something could be so closely tied
Up with our bodies if it slithered in from the outside,
690 When observation shows us quite the opposite is true;
For the spirit is so closely bound up with the veins and tissue,
Sinews and bones, that even teeth possess a share of sense:
We know from toothaches, and the way ice-water makes us
        wince,
Or chomping on a piece of hard grit in a chunk of bread.
Since spirits are so closely woven with flesh, it's clear instead
They are not able to extract themselves all in one piece
Nor safely from all sinews, bones and joints obtain release.

But if you happen to think it is the spirit's habit to slide
Into our frame and seep into our limbs from the outside,

Then it is that much likelier to perish, being alloyed
With flesh; what permeates, dissolves, and therefore is
    destroyed.   700
For the spirit, distributed through every limb and organ, sent
Down every passageway, is broken down like nourishment,
Yielding another substance altogether. Thus the soul –
Mind and spirit – even if it enters a newborn whole,
Dissolves by permeating; it must scatter to enmesh,
Spreading its particles through every passage of the flesh.
And so the particles that make this mind, the very same
That holds sway over the body now, rise from the mind that
    came
Into the flesh and passed away, all parcelled through the frame.   710
Therefore, it is evident the spirit has a day
When it is born, and has an hour when it must pass away.

Another thing – are particles of spirit left behind
Inside a lifeless corpse, or not? Because if we should find
They *are*, we have no rightful grounds on which to hold it true
That spirit is immortal, since it was, as it withdrew,
Diminished by lost fragments; but, if it escaped instead
With all its parts intact so that it did not leave a shred
Of itself inside the body, then explain how corpses breathe
Forth maggots from their putrifying flesh? Where does the
    seethe   720
Of bloodless boneless creatures come from that goes swarming
    through
The bloated limbs? However, if by chance you think it's true
That spirits are possessed of the ability to squirm
From the outside, one by one, into the body of each worm
(If you don't think it queer thousands of spirits should
    convene
Exactly at the spot where only one has quit the scene),
A line of questioning that it might be worthwhile to take
Is whether spirits hunt out seeds of tiny worms, and make
Themselves a dwelling out of these in which they can abide;
Or do they find worms ready-made and fit themselves inside?

730 But why in the world would spirits bother, why would they take
        pains?
     For while they have no bodies, they flit about free from the
        banes
     Of Sickness, Cold and Hunger; it is, rather, *flesh* that strains,
     Liable to these failings; and the mind is also prone –
     From such close contact with the flesh – to ailments of its own.
     Granted, for spirits it would be convenient to construct
     Bodies to enter – but there's a vital question we have ducked:
     How could they possibly do it? Spirits therefore do not make
     Limbs and bodies for themselves, nor, clearly, do they snake
     Into completed bodies either, since they could not then
740 Combine to forge the intricate links of mutual sensation.

     Indeed, why does bitter ferocity go with the louring breed
     Of lions, and craftiness with foxes? Why is flighty speed
     Bequeathed deer from their sires, their legs spurred by ancestral
        dread?
     And all the other qualities of this type, why are they bred
     In the flesh and character right from life's start, unless the mind
     And its powers grow along with the flesh according to seed and
        kind?
     But if the mind were deathless, and it were its wont to change
     One body for another, creatures would start acting strange,
     Confused in their behaviour: the attack of antlered deer
750 Would make Hyrcanian hounds[17] turn tail and bolt away in fear;
     The hawk would quail mid-air, fleeing the dove in hot pursuit;
     Mankind would lack all logic, reason crown the savage brute.

     For if you think that an immortal spirit changes along
     With the alteration of the flesh, your reasoning is wrong:
     To change is to dissolve, and to dissolve, to pass away.
     The spirit's parts are rearranged, and from their order stray;
     Therefore they must be able to dissolve throughout the frame,
     And finally to perish with the body just the same.

760 But if some would insist a human spirit always flies
     To a human body, how do foolish spirits come from wise?

Why was there never a child with any forethought, nor a foal
That knew its paces like a grown horse trained in *haute école*?
They'll wriggle out of it, by claiming that the mind will grow
Enfeebled in a feeble body. But even were that so,
They must confess the mind is mortal, since it's changed all through
    through
The flesh, and lost the life and sense that it had hitherto.

How can the mind strengthen with the flesh at the same time,
How can it attain the yearned-for flower of its prime,   770
Unless it was its comrade from the first, the very start?
Or why, from agèd limbs, does it desire to depart?
Is being shut up in a rotting corpse the thing it fears –
Its dwelling-place, dilapidated with the lapse of years,
Come crashing down? But for the immortal, there can be no
    dangers.

And it's ridiculous to think that as wild beasts are mating
And whelping they're surrounded by immortal spirits waiting
For mortal bodies to move into, mobbing myriads deep,
Jostling and elbowing to be the first to creep   780
Inside the flesh. That is, unless instead the spirits keep
Some deal they've struck amongst themselves – so that the first
    to light
Upon a body gets to slither in without a fight.

Again, a tree can't live up in the sky, a cloud can't scud
In the deep sea, a fish can't live out in the fields, nor blood
Pulse through bark and branches, nor does sap flow through a
    stone,
But each thing has a fixed place where it grows to call its own.
The nature of the mind cannot arise, sans flesh, alone,
Nor is it able far from blood and sinew to abide.
But even if it could, the mind would likelier reside   790
In the head, or in the shoulders, or the very heels of the feet,
Or born in any part you please, just so it keep its seat
In the same person, the same vessel. But since there is assigned
Even *within* our bodies a fixed and separate place for mind

And spirit to grow and dwell, all the more reason to deride
The thought they could arise and could endure somehow
     outside.
Thus when the body is destroyed, you must admit the soul
Passes away, shredded through the body as a whole.
800 Indeed, to harness mortal and immortal with one yoke
And think they can agree, and interact, is but a joke.
For what could be more out of tune, off-kilter and contrasting
Than a mortal thing that's hitched to something deathless,
     everlasting,
To weather with their wedding tempests furious and blasting?

Besides, there are three types of things that last forever: those
That being utterly solid in their substance shrug off blows
And do not permit anything to penetrate inside
The close-knit fabric of their parts, to rend them and divide
810 (Atoms are of this type, as I have shown you a while back);
Or else the reason things can last forever is they lack
Anything to *do* with blows, such as the Void, for it
Remains untouched by blows, and blows affect it not a bit;
Or else it is because there is around them no supply
Of space into which their parts are able to dissolve and fly,
Such as the Universe, which is eternal, since there's no
Place beyond it where its scattered elements could go,
No bodies that could fall on it and crack it with a blow.

But if you think the spirit is immortal, on the ground
820 That it is kept by vital powers sheltered safe and sound,
Or else that no threats come along at all that would attack
Its well-being – or if they come, retreat, somehow pushed back
Before we're able to perceive what mischief they can do,
[Observation shows us that these arguments aren't true]
It's not just that the spirit, when the flesh ails, falls ill too,
But often it's tormented by what hasn't happened yet,
Sick with dread, worn out with fretting; or gnawed at with
     regret
For sins that were committed sometime in the past. And add
Ailments that are only of the mind, like going mad,

Or losing the memory of things. And add that it sinks down
Whelmed beneath the pitchy waters of Oblivion.

Then Death is nothing to us; it concerns us not a jot,   830
Seeing we hold the mind is mortal. And just as we did not,
In time gone by, feel anxious when the Carthaginian host
Swarmed into the fray from every quarter, every coast;
And the whole world – everything beneath the sweeping shore
Of heaven – trembled, shaken by the sickening shock of war,[18]
And when on land and sea the rule of all of Mankind lay
In the balance, which of two empires was destined to hold
     sway,
So when the bond is put asunder between body and soul,
The two from which we are composed into a single whole,
Nothing can befall us, we who shall no longer *be*,   840
Nor move our senses, no, not even if the earth and sea
Were confounded with one another, and the sea mixed with the
     sky.

And let's say even after they were ripped out of our body
The mind and spirit *could* feel. Well, so what? That does not
     mean
Anything to us, a cobbled coupling between
The spirit and the flesh. And even if time *could* somehow
Gather our atoms together after death as they are now,
And we were blessed with the light of life again – it wouldn't
     matter
Anything to us, not once our recollections scatter   850
That connect us with our former selves. And what we were
     before
Is no concern to us as we are now, nor any more
Are we haunted by their former sufferings. For when you cast
Your gaze back over the whole immeasurable span of ages
     past,
And consider matter's motions, a diversity so vast,
It's easy to believe that these same seeds of which we're made
In the here and now have in the past been frequently arrayed

In the same way, although we can't remember it because
860 A caesura has been cast between those different lives, a pause,
And the motions of their consciousness have wandered far
     astray.

For if someone will ail and suffer at some future day,
He must *exist* in that time when the maladies beset.
But Death removes the possibility, since Death won't let
The man exist for whom these ills are hoarded up. It's clear,
Therefore, that Death is absolutely nothing we need fear,
And that he who is *not* cannot be wretched or forlorn.
What can it matter to the man that he was even born
Once Deathless Death despoils him and his mortal life is shorn?

870 So when you see a fellow who is outraged with his lot –
That after death his corpse must either be entombed and rot,
Or be devoured by flames, or by the jaws of savage beasts,[19]
The note he sounds is false – some hidden goad lurks in his
     breast.
Though he denies believing there's sensation after death,
He contradicts his premise; his protests are a waste of breath –
He does not rip himself up by the roots away from life,
But rather, unwittingly, has something of himself survive.
For when someone alive imagines that his flesh will be
880 Torn at death by vultures and wild animals, his pity
Is for *himself*, for he does not sufficiently divide
Himself from *it* – the cast-out corpse – and standing at its side
Thinks it is he, tainting it with his feelings. That is why
He resents he was created mortal, for he does not descry
That in true death, no part of him will stay alive to mourn,
Standing by to suffer as he lies on fire or torn.
For if, in death, to be savaged by the fangs of beasts is dire,
890 I fail to see how it's more pleasant on the scorching pyre,
Roasted in flames, or in embalming honey to be drowned,
Or to freeze stiff on a slab of icy stone, or have a mound
Of earth crushing and weighing on you six feet underground.

'No more happy welcome-home, no waiting wife to miss you,
No pitter-patter of little feet as children race to kiss you,[20]
Touching your heart with wordless tenderness. Alas, no more
Can you provide for them, you can't keep danger from their
*   door.*
Unlucky man! One dark day snatched these joys of life from
*   you,'*
They cry, but do not add, *'and all the yearning for them too.'*   900
If they could see it clearly, and their words followed in kind,
They'd free themselves from heavy dread and anguish of the
    mind.

'And as for you there, slumbering in Death, you shall be free
From pain and grief for the remainder of eternity.
But we, hard by, as you lie charred to ashes on the heap
Of the dreadful pyre, can't slake our grief however much we
*   weep;*
No day shall ever lift eternal mourning from our breast.'
We should ask the man who says such things: if it all ends in rest   910
And sleep, what is so very bitter that anyone could pine
Away in everlasting grief?

                          And often, when men recline
At table, shady wreaths upon their brows, they lift a glass
And utter from the bottom of their hearts, 'How swiftly pass
The pleasures for us poor old mortals! They are gone before
You know it, and they cannot ever come back any more.'
As though in death, of all the evils, the one that will be worst
Will be that they are parched and tortured with a searing
    thirst! –
Or else some other hankering. In fact, no one regrets
The loss of himself or his life when both the mind and body rest   920
In sleep. As far as we're concerned, this sleep could last forever
Because no longing for ourselves afflicts us. Yet in slumber
Those atoms all throughout our limbs don't wander far and
    stray
From sense-producing motions, since a man who's snatched
    away

From sleep can pull himself together. So Death should mean
    much less
(If something can be 'less' than what we see is nothingness),
For there's an even greater scattering of matter shaken
At death. And once the chill standstill of life has overtaken
930 Someone, he will never rise again nor will he waken.

And then, what if Nature herself suddenly should ask
These words, lifting her voice, taking one of us to task:
'What's so much the matter with you, mortal, that you wallow
In morbid mourning? Why bemoan your death and weep in
    sorrow?
For if you've relished the life that you have led, if you did not
Gather all your blessings, as it were, in a leaky pot
So that they've drained away and perished, with no chance to
    please,
Why not, like a banquet guest, who's drunk life to the lees,
Depart, you dolt, and go to peaceful rest, your mind at ease?
940 *But* if all the good you got was wasted, poured away,
And life is hateful to you, why seek to extend its stay? –
All will just turn out wrong and perish profitless again.
Why do you not, instead, make an end of life and all its pain?
For there's no further pleasure I can think up or invent
For you – it's always the same. And even if your limbs aren't
    spent
Already and feeble, and even if the years don't waste your
    frame –
Even if you outlived the generations, and you became
Immortal – even then, it would still be *more of the same*.'
950 And what can we say, except that Nature argues on just ground,
And that the case she makes against us is completely sound.

But if one getting on in years should grumble, and deplore
His own demise pathetically, would not She all the more
Rightly exclaim against him, and denounce in sharp tones:
'Away with your tears, you bottomless pit, stifle your whines
    and moans!

For having had as much of Life's rewards as you could reap,
You wither, but since you lust for what you don't have, and hold
    cheap
What is at hand, Life, wasted and unfulfilled, slipped through
    your fingers.
Death has snuck up on you, now at the head of your couch he
    lingers
Before you're ready to depart Life's banquet, full and sated.   960
Now put aside all thoughts that are unseemly for your grey
    head,
Come along, make room for others, leave with your heart light.
So it must be.'

                And I think that She would be in the right,
And right to carp and chide. The old must give way, pushed
    aside
By the new, and one thing by another thing is re-supplied.
None's consigned to the pit, to pitch-black Tartarus, below –
Future generations need material to grow.
And they, when life is through, shall follow you into the grave,
As those that came before, no less than you, wave after wave.
Thus one thing rises from another – it will never cease.   970
No one is given life to own; we all hold but a lease.
Look back again – how the endless ages of time come to pass
Before our birth are nothing to us. This is a looking glass
Nature holds up for us in which we see the time to come
After we finally die. What is it there that looks so fearsome?
What's so tragic? Isn't it more peaceful than any sleep?

And certainly whatever things are rumoured to dwell deep
In Acheron are all things that exist in life for us.
There doesn't exist, as in the tale, a pitiful Tantalus[21]
Who stands afraid, frozen to the spot from empty dread
Beneath a huge rock dangling in the air above his head.   980
But, rather, in life it's vain fear of the gods that hangs above
Mortals and downfalls of Chance they live in terror of.

There is no Tityus prostrate in Hell, who's ripped apart
Forever by flapping vultures. Nor when they probe his giant
    heart
Is there sufficient sustenance for them on which to dine
Throughout eternity. No matter how huge, lying supine,
His body is, his limbs sprawled out not only over nine
Acres, but extended across the earth's entire sphere –
990 He won't be able to endure the agony forever,
Nor can he offer flesh to make them endless dinners of.
But Tityus is here with us: stretched on the rack of Love,
As he's pecked by flapping cupids, eaten up with jealous care –
Or the worries of some other passion rip at him and tear.

And Sisyphus exists in life, right here before our eyes:
The man consumed with seeking the accoutrements of office
From the people, who always comes back sad and beaten. To be
    driven
To seek power – an illusion after all – which is never given,
And undergo endless hard toil in striving for it still,
1000 This is the act of struggling to shove a stone uphill,
Which, at the very peak, only goes bounding down again,
Seeking, quick as it can, the level field of the campaign.

And then, always to feed a mind that won't be satisfied,
And not to fill it with good things however much you've tried,
As the different seasons satiate us in their yearly round
With all the different fruits and pleasures in which they abound
(Though when it comes to the fruits of life, we always have more
    room) –
This is where we get the tale of maidens in their bloom,[22]
Who, as the story goes, pour water in a leaking pot,
1010 And no matter how they try to fill it up, still they cannot.

And Cerberus, and the Furies, and the dearth of light as well,
And noxious vapours belching forth out of the maws of Hell –
These do not anywhere exist, nor can they, it is clear.
Rather, it's punishment *in life* for misdeeds that we fear,

In proportion to the dreadful deed, and how we must atone
For crimes committed: locked away in a prison cell, or thrown
From Traitor's Rock;[23] the cat-o'-nine-tails, torturers, the rack,
Pitch, red-hot metal plates, firebrands – for even though we lack
Such things, the Conscience fears for its misdeeds, and *it* applies
Goads and floggings to *itself*, neither does it surmise,
Meanwhile, that there can be, to all its sufferings, an end,   1020
Nor, at length, a final limit to its punishment,
So it dreads these same oppressions will grow weightier with
    death,
Until, at last, the life of fools becomes a Hell on Earth.

You might, from time to time, give yourself this to recite:
'*Even Ancus the Good has looked his last upon the light,*[24]
Who was a better man than *you* by far, you reprobate,
And since his day, the sun of many a king and potentate
Who held sway over mighty peoples has set. Yes, even he[25]
Who for his legions paved a road across the great blue sea   1030
And taught them how to stride the salty main, he who held
    cheap
The ocean's roar and with his horses trampled on the deep –
Robbed of light, his spirit fled, he too went to the grave.
And Scipio,[26] firebrand of war, the Scourge of Carthage, gave
His bones unto the earth like any slave of humble duty.
Add to these the pioneers of Wisdom and of Beauty,
Add the companions of the Muses, poets of renown.
Even Homer, the one and only who deserves the crown,
Even he now sleeps one sleep with all the rest. The sage
Democritus, when he was warned by his advanced old age
That the motions of his mind – his very memory – were fading,   1040
He *himself* gave his own head to Death, unhesitating!
Even great Epicurus,[27] once the light of life had run
Its course, perished, the very man whose brilliance outshone
The human race, eclipsing all, just as the burning sun,
Risen, snuffs out all the stars. So who are *you* to balk
And whine at death? You're almost dead in *life*, although you
    walk

And breathe. You fritter away most of your time asleep. You
    snore
With your eyes open; you never leave off dreaming, and a score
Of empty nightmares fills your mind and shakes it to the core.
1050 Often, addled and dizzy, you don't even know what's wrong –
You find yourself besieged at every turn by a whole throng
Of cares, and drift on shifting currents of uncertainty.'

Men feel a heaviness upon their minds, it's plain to see,
That weighs them down. If they could grasp the cause of this
    ennui,
This heap of misery and care that hunkers on the heart,
They would not lead the lives we see they *do* for the most part,
None knowing what he wants, each ever seeking a change of
    place –
As if he could lay his burden down by travelling through space.
1060 Often a man who's sick and tired of his own hearth will roam
From his roomy mansion, only to come suddenly back home
Because he feels no better when he's somewhere else. He heads
For his country villa, driving his imported thoroughbreds
Hell-for-leather, as though to save a house on fire. And yet
The fellow starts to yawn the very moment he has set
Foot in the door, or falls in a heavy sleep, seeking to drown
In oblivion. Or even wants to hotfoot back to town!

Thus in this way each man is running from himself, yet still
Because he clings to that same self, although against his will,
1070 And clearly can't escape from it, he loathes it; for he's ill
But doesn't grasp the cause of his disease. Could he but see
This clear enough, a man would drop everything else, and study
First to understand the Nature of Things, for his own sake:
It's his condition for *all time* – not for one hour – at stake,
The state in which all mortals should expect themselves to be
After death, for the remainder of eternity.

For what's this great and wicked lust for living all about,
If it just drives us to distraction, amidst danger and doubt?

The life of mortals has a limit set to it, my friend.
Death has no loopholes. All of us must meet it in the end.
We go through the same motions in the same old place. No
    measure   1080
Of added life will ever coin for us a novel pleasure.
True, while we lack that which we long for, it is an obsession,
But we will just crave something *else* once it's in our possession;
We are forever panting with an unquenched thirst for life.
No one knows what the years to come will bring – what joy or
    strife
May lie in store for us, what outcome's looming in our lot.
But by adding on to life, we don't diminish by one jot
The length of death, nor are we able to subtract instead
Anything to abbreviate the time that we are dead.

Though you outlive as many generations as you will,   1090
Nevertheless, Eternal Death is waiting for you still.
It is no shorter, that eternity that lies in store
For the man who with the setting sun *today* will rise no more,
Than for the man whose sun has set months, even years, before.

BOOK IV

THE SENSES

I wander in the uncharted country of the Muses – none
Before me has set foot here – and I thrill to come upon
Springs untouched by any lips, and here to slake my thirst.
I joy to pluck strange flowers for a glorious wreath, the first
Whose brow the Muses ever crowned with blossoms from this
    spot.
Why? Because I teach great truths, and set out to unknot
The mind from tight strictures of religion, and since I write
Of so darkling a subject in a poetry so bright,
Tingeing with the Muse's grace all subjects that I touch.
10 Nor is my method to no purpose – doctors do as much;
Consider a physician with a child who will not sip
A disgusting dose of wormwood: first, he coats the goblet's lip
All round with honey's sweet blond stickiness, that way to lure
Gullible youth to taste it, and to drain the bitter cure;
The child's duped but not cheated – rather, put back in the
    pink –
That's what I do. Since those who've never tasted of it think
This philosophy's a bitter pill to swallow, and the throng
20 Recoils from it, I want to coat this physic in rich song,
To kiss it, as it were, with the sweet honey of the Muse,
So I might keep you focused on my verses by that ruse,
While you drink in the Nature of Things, and know it of good
    use.

Already I've explained to you the nature of the mind
And by what means it thrives and grows in strength when it's
    combined

With the flesh, and how, once sundered from the body, it will
    scatter
Reduced to basic elements. Now there's another matter
Of vital importance that I must explain. Let me begin
By saying there are images of things – a sort of skin   30
Shed from the surfaces of objects, from the outer layer –
Films that drift about this way and that upon the air.
And it is, rather, *images* of things, these films, that make
Our minds afraid when we encounter them while we're awake,
And these are what we see when we are dreaming and catch
    sight
Of the weird and eerie effigies of those Lost to the Light,
And which, when we lie sunk in slumber, rouse us with a
    fright;
So let us not imagine ghosts have broken out of Hell,
And here above, among the living, flitting phantoms dwell,
Or in the wake of death, that any part of us abides
Once soul and body have parted company, and on both sides   40
Perished, each reduced to its own atoms. I say, therefore,
An object gives off likenesses from its exterior,
The flimsy shapes of things. And this I shall proceed to show
So anyone can grasp, however dim he is or slow.

[I've taught you what the nature of the atom is, and I
Have shown they come in different shapes; I've taught you that
    they fly
About at will, goaded into never-ceasing motion,
And how these atoms come to make up all things in creation.
Now there's another crucial fact I must explain – so mark
My words – that there are images of things – a skin, or *bark*,   50
As we can call it, shed from objects, since it bears the same
Form and likeness of whatever thing from which it came.][1]

First of all, since we see many objects give off matter
Right out in plain view – sometimes loosely in a scatter,
As burning wood sheds smoke or as a fire gives off heat,
Or sometimes shedding substances more tight-woven and neat –

As commonly cicadas doff the shiny shirts they've worn
In summertime, or when calves shed the birth-sac as they're
        born,
60 Or the slippery serpent shrugs her castings off upon the thorn
(For often we behold such spoils a-flutter on a briar);
Since these examples come to pass, it follows that there are
Flimsy images as well shed from the outer layer
Of things. For if skins peel and drop away before our eyes,
Why not thinner films? No one could hint it's otherwise –
Seeing that there are many tiny particles on the face
Of things that can be shed while their arrangement stays in
        place
So they retain the stamp of that thing whence they came;
        indeed
70 Because these films are on the surface, little can impede
Their progress, and their fewer particles give them greater speed.

For clearly we see many things emit bodies galore,
Not just from deep inside themselves, as I have said before,
But also from exteriors, for instance with their hue –
As red and gold and purple awnings regularly do,
When, stretched over a great theatre, hung on poles and beams
Above the crowd, they flap and billow, and their colour streams,
Staining the faces of the audience in the stands below,
80 The stage and spectacle a-flicker in their fluttering glow.[2]
The more the theatre's enclosed, shut from the glare of day,
The more the laughing show of colours puts on its display.
Therefore, since linen, from its outer surface, gives a hue,
So everything must be emitting flimsy images too –
Since, as with colour, these are thrown off from the outer
        surface.
There are such things as films, therefore, that keep the certain
        trace
Of forms, flying everywhere, of such a gauzy weave
That separately, one by one, they're impossible to perceive.

90 And the reason things like heat or smoke or odours overflow
And spill forth out of objects, scattering abroad as they go,

Is they arise from deep inside a thing, and so are rent
To pieces as they exit by paths tortuous and bent,
For there is no direct way out to all push through together.
On the other hand, when thin slips of a superficial colour
Are cast off, there is nothing that can rip them up or tear,
Since they are right out on the surface, exposed to open air.

Lastly, likenesses we see reflected in a glass,[3]
In water or in shiny surfaces must come to pass
(Because in their appearances they are an object's twin)   100
From images shed by the thing itself. And therefore thin
Shapes of things and likenesses exist, which although none
Is able to detect them as they come off one by one,
Yet when they bounce back thick and fast in a continuous flow
From the flat face of a mirror, form the images that show.
How images in a glass are able exactly to retain
An object's form, there is no other good way to explain.

Come now, and learn how thin the texture of the image is: 110
For one thing, atoms are beneath detection of the senses,
Much tinier than the point at which our eyes fail to perceive.
And I shall demonstrate this briefly here, so you'll believe
How fine-spun are the warp and woof of things, all matter's
     weave.

First, some animalcules are so minute, you couldn't see
Anything at all if you chopped one up into three.
How are we to grasp how unimaginably small
Its guts must be, or the pinpoint of its heart, or its eye's ball?
Its members and its joints – how small are they? As for the size 120
Of the particles of its mind and spirit – don't you realize
How tiny and how delicate they'd have to be?

                                        As well,
Take anything that breathes forth from its flesh a pungent
     smell –
Heal-all, loathsome wormwood, overpowering lad's-love,[4] fell

Centaury – and then lightly pinch it with your thumb and finger:
[It leaves its odour clinging, but the particles that linger
On skin, are such that you cannot detect them with your eyes.]
So there are also images – this you should realize –
That have no impact on the senses, floating about likewise.

But if you think the only images that drift along
130 Are images that peel away from objects, you are wrong;
For there exist, as well, those that spontaneously arise,
Forming on their own up in the air high in the skies
And drifting about, just as we sometimes see clouds mass
        together,
Mounting rapidly on high and frowning on fair weather,
Their motion ruffling the air. For frequently clouds make
Giant floating faces, trailing long shadows in their wake.
At other times huge peaks or boulders wrenched from mountain
        crags
140 Drift over and across the sun, while next some Monster drags
More thunderheads behind it; thus formations are forever
Melting and shifting, taking the shape of anything whatever.

Now then, how fast and easily these images come about,
Ever streaming from things, and gliding away [I shall set out.]
For something's always overflowing from the outer surface
Of objects, which they then throw off – something which can
        pass
Through certain substances it meets, just as it goes through
        glass;
Yet when it meets with rough stone or with solid wood, will
        crack;
It shatters at that instant, and can give no image back.
150 But on the other hand, when it meets with an obstacle
That is both polished and tightly knit, the mirror most of all,
Nothing like this happens. For although it cannot pass,
As through glass, neither does it shatter. The polished face
Of the mirror is ever mindful to guarantee its safety. Thus
It turns out that the likenesses flow back, from it, to us.

Thrust an object quickly as you can before a mirror –
Its image is there. From this experiment, what could be
      clearer? –
These gauzy films and flimsy shapes continually pour
From the surfaces of objects. Many likenesses, therefore,
Are generated in so short a span of time, the speed   160
Of their generation is a fact that we have to concede.
And as the sun must send so many particles of light
Down in such swift succession it can make all places bright,
And without any interruption, flood them with its beams,
So things at any given moment also give off streams
Of images in all directions – since wherever we place
A mirror, whatever angle to an object it may face,
It gives a picture back that corresponds in form and hue.

Besides, consider how, of a sudden, out of the clear blue,
The heavens go so black and threatening that you might well
Imagine all the darkness had been emptied out of Hell   170
From every quarter to shroud the great celestial vault with
      gloom,
Such a fell night of clouds amasses, while overhead there loom
Faces of black horror in the sky. And yet how small
And slight an image is compared to clouds, no one at all
Can say, or even give a clear idea.

                              Now we shall see
How swift these likenesses zip along, how easily and free
They can go swimming through the air – moving at such a pace
It takes them hardly any time to cross vast spans of space,
Each heading to its separate destination. I shall treat
This topic in my verses, but I'll make it short and sweet –   180
For the brief song of the swan is better than the squawking cries
Of a far-flung flock of cranes up in the clouds of southern skies.

First, we notice frequently that objects that arise
From very fine and tiny elements are also fleet.
Consider, for example, the light of the sun and the sun's heat:

Because the particles that they consist of are minute,
And are pummelled, as it were, they do not hesitate, but shoot
Through the air that lies between, each driven on by blows that
    smite
From the one behind, so instantly light's on the heels of light,
190 And flash prods forward flash as if in one continuous team.[5]
Thus images must be able, by the same process, to beam
Across an untold space in hardly any time at all;
First, because there is a push from far behind, though small,
That launches and propels them onward; next, because they're
    light
And travel with a corresponding nimbleness of flight;
Lastly, they're cast off in such loose texture they can drift
Easily through anything, and as it were can sift
Through all the air that lies between.

                             Furthermore, consider
200 Those bodies things emit from deep inside themselves and
    scatter
Abroad, such as the sun's heat and its light – if these can fly
So that we see them sweep across a vast expanse of sky
At the moment the day breaks, and wing across the sea and land
And flood the heavens, then what of those that lie ready to hand
On the very surface of things, where there is nothing to impede
Their launch? For don't you realize they must have greater
    speed
And travel farther, crossing space much vaster in extent,
In the same time that it takes sunlight to crowd the firmament?

And here's an indication that's particularly clear
210 Of the great velocity with which these likenesses career –
When you set a bowl of water underneath fair skies at night,
Instantly it twinkles back with heaven's starry light.
Now don't you see how images from the rim of heaven fall
To the borders of the earth in scarcely any time at all?
Therefore, I say again and again, you must admit [the speed
At which the image travels is] miraculous indeed.

[...]
And there are particles which smite the eyes and make us see,
And odours pouring forth from certain things perpetually
As coolness flows from rivers, heat from the sun, and as a spray
Of brine pours from the ocean waves and gnaws shore walls
    away.   220
Different sounds go flying through the air and never halt.
And often strolling by the sea, we get a tang of salt
In the mouth, or when we stand by and we watch someone
    prepare
A dose of wormwood, we can taste a bitterness in the air.
Yes, clearly many different particles are streaming away
From all things, strewn in every direction, and there is no delay
Or respite to interrupt this flow, since we perceive an object
Without interruption, and at any time we can detect
A thing by means of sight or smell or listening to its sound.

Besides, since shapes we know by handling in the dark are found   230
To be the same seen in the light of day, then sight and touch
Have to be triggered by similar causes. Now then, if we clutch
Something that's square, and it sets off our senses in the night,
What square thing can be met with by our vision in the light,
If not the image of that thing? Therefore what causes sight
Is images, which nothing can be visible without.

These films which I call likenesses of things are scattered about
And given off on every side, in every direction.   240
But since our sight's the only sense equipped for their detection,
It happens that wherever we turn our eyes, wherever we view,
Everything comes to meet our vision with its shape and hue.

It is thanks to the image that we understand how far
Away from any object we're observing that we are.
Launched, it instantly shoves before it all the air that lies
Between the object of origination and our eyes,
So all this air comes flowing through our eyeballs, and the wind
Rubs the edges of the pupil as it crosses in.

250 And that's how we judge distances; the more air in the blast
Pushed before an image, the longer that the breeze will last
Scrubbing against the edges of our pupils as it streams,
And thus the farther away removed from us the object seems.
Doubtless, this happens in such a very rapid fashion, we
Perceive the object and its distance simultaneously.

And on the subject of vision, it should come as no surprise
That rather than seeing every single film that strikes the eyes,
We perceive the very object itself. For although the wind blows,
Whipping us with many lashes, and the fierce cold flows,
260 We are not wont to feel each cold and windy particle
One by one but, rather, feel the blast *all as a whole*,
And we see these blows affect our flesh as if an object jars
And gives a sense of its own body there outside of ours.
And then consider when we stub a toe upon a rock,
How it is just the outmost surface of the stone we knock,
The topmost layer of colour, yet that isn't what we feel;
Instead we sense the inner hardness that its depths conceal.

Now let's consider why it is an image seems to sit
270 Behind the surface of a mirror, deep inside of it,
Since that is what we see. The answer is that it's the same
As with those objects that we actually glimpse beyond the frame
Of an opened door, when the door affords us an unobstructed
        view
Of things beyond the house's walls. For this is also due
To double blasts of air, since it is first the air that lies
Between us and the doorjambs that comes blowing through our
        eyes,
And then the image of the doors themselves, left leaf and right,
And next the outdoor light and outdoor air brushes our sight,
Then all those things we rightly see beyond the doors, outside.
So when a mirror's image is first cast off, as it flies
280 To our pupils, it pushes all the air between it and our eyes,
And makes us feel all this before we see the glass. But when
We see the looking-glass itself as well, instantly then

The image that we cast ourselves arrives and hits the glass
And bouncing, strikes our eyes again, and rolls another mass
Of air in front of it, and makes us feel the air-blast sweep
Before our image: that's why it seems set in the glass so deep.

Therefore I say again, it shouldn't surprise you any more
[That the same happens when we really see things through a
     door]
As when we look at things reflected from a mirror's face,   290
Since this is due to double waves of air in either case.

Now the reason in a mirror that the body's right-hand side
Appears on the left, is that, when on-coming images collide
With the flat plane of the mirror, they are not safely turned
     around,
But rather, they are dashed directly backwards and rebound;
Just as if someone took, before it dried, a mask of clay
And dashed it against a beam or post, at once its shape would
     stay
Just as it was in front, but pressed out in relief behind,
So the eye, once on the right, is on the left now, and we find   300
The left eye on the right – they have switched places with each
     other.[6]

An image can be handed from one mirror to another
So that as many as half-a-dozen images appear.
Consider something hidden back in the interior
Of a house; however deep inside, however convolute
Its path, its image can be led outside along a route
Of several different mirrors in a series, down the halls,
Bending round the corners, and be seen outside the walls,
So faithfully from one mirror to another it reflects.
And that which faces left, flips to the right again, and next
Flips back to what it was.

                              Or take a mirror that's endowed   310
With two curving sides, bowed like our own ribcage is bowed:

The image of ourselves that such a mirror gives to us
Is not reversed – right corresponds with right – either because
The image is tossed between the mirrored surfaces, and flies
From there to us, dashed back and forth and back, inverted
    twice,
Or else because the image is wheeled around just as it reaches
The mirror, and turns towards us as the mirror's curved shape
    teaches.[7]

And then, our likeness walks in lockstep with us, placing its feet
320 As we do, aping our every move, because when we retreat
From any part of the mirror, at once the image from that place
Can't be reflected any longer from the mirror's face,
Since Nature makes it that all images return, bounced back
Along an angle equal to the angle of attack.[8]

The eye, moreover, turns away from bright lights. You will find
If you look upon it long enough, the sun will make you blind –
Because of the sun's own power, and because its images bear
Down from such a towering height, plummeting through clear
    air,
Smiting the eyes and throwing their composition all awry.
Moreover, what is blindingly bright will often sear the eye
330 Because of the many seeds of fire that it must contain,
And these seeds, as they pierce inside the eyeball, cause it pain.

Furthermore, all things that sufferers of jaundice view
Become a sickly yellow, since seeds of this lurid hue
Pour forth from their flesh and run into on-coming films, and
    too,
Since many seeds are mixed within their eyes, so that at last
By contact, everything is tainted with that sallow cast.

The reason why we see things out in sunlight when we stand
In darkness is that when the murky air that's closer to hand
Enters and occupies the open eyes, without delay
340 The bright and shining air comes after, washing it away

And scattering all the gloomy shadows of the air before;
For shining air is made of many particles far more
Nimble, minute and powerful. As soon as air that's bright
Has flooded full the pathways of the eyes again with light,
And opened up what had been under siege by the black air,
In a trice, the images of things in light come pouring after,
And they then stimulate our sight. But on the other hand,
We can't make out what's lurking in the darkness when we
      stand
In light, because, after the light, the gloomy air invades,
A coarser air that clogs the portals up and that blockades   350
The pathways of the eyes, so that the images that smite
The eyeballs are not able then to stimulate the sight.

And why four-sided towers of a town are often found,
When looked at from a distance, to appear not square but
      round,
Is that discerned from far away, sharp corners become dull,
Or rather, their brunt is lost and they become invisible,
Their impact never reaching our eyes at all, because the mass
Of air the images must cross buffets them as they pass
And blunts them. Therefore, when no corners reach our eyes
      unscathed,   360
The masonry seems polished smooth, and rounded as if lathed –
Not rounded like a close-up thing that's *really* round, it's true,
And yet there is a sketchy resemblance between the two.

We also notice the movement of our shadows in the sun
Following in our footsteps, mimicking gestures, every one.
(That is, if you believe that air deprived of light can take
A stride, or follow the gestures and the movements that men
      make.
For those things commonly called 'shadows' – what else can they
      be
If not air that's devoid of light?) And indisputably
This comes about because the ground's robbed of the light of
      day   370
In one part, then another, as we wander in its way,

And when we leave, that area fills again with the sun's glow
So that it seems one shadow follows us wherever we go.
New rays of light are ever pouring forth, while old expire,
Just like a thread of wool that's being spun into a fire.[9]
Thus earth is easily despoiled of light, and easily made
Full of light again, rinsing away the gloomy shade.

But I do not allow the *eyes* are tricked in this at all.
380 Their task is only to discern where light and shadows fall;
Whether the light's the same or not, or the shadow that was *here*
Is one and the same with the shadow that's now passing over
    there
(Or whether this effect occurs as I just now defined),
Nothing can determine save the reason of the mind.
Eyes can't grasp the true nature of things. So do not claim
The fault's with them, when really it's the mind that is to blame.

The scudding ship on which we sail seems to be standing fast,
While yet a craft at anchor will appear to sail on past.
And it is, rather, hills and fields that seem sternwards to fly
390 When really the ship is rowing, or under sail goes skimming by.
The stars all seem to be at rest, nailed in the vaults of heaven,
But are perpetually in motion, since once they have arisen,
Returning to their setting place, on far-flung paths they go
Across the sky from end to end, with bodies all aglow.
The sun and moon seem likewise to be frozen still, but prove
On observation, actually, to both be on the move.
From far off, mountains jutting from the middle of the sea
With space enough between for fleets to pass through easily,
Seem nevertheless to be linked up into a single isle.
400 Dizzy children think the columns in a peristyle
Are going round, and that the entire court is in a spin,
Once they themselves stop turning, so that they almost begin
To believe the house threatens to tumble in about their ears.
And then, when first above the tops of mountains Nature rears
The beaming sun all red with flickering flames, the sun appears
To be so close it's singeing them with fire. Those mountains are
Perhaps two thousand arrow-shots away from us – not far –

Maybe even as close as a mere five hundred javelin-casts.
And yet between the mountains and the sun there lies a vast   410
Expanse of ocean strewn beneath the sweeping shores of
    heaven,
And also myriads of countries stretching in between
Peopled by the sundry tribes of animals and men.

And yet puddles of water that upon a paved street linger
Between the cobblestones, although no deeper than a finger,[10]
Offer us a downward glimpse into the earth as deep
As the yawn of heaven overhead is towering and steep,
So that it seems you're peering down into the cloudy skies,
Or you behold a moon and stars – you can't believe your eyes –
Buried underneath the earth. Then, when a headstrong horse   420
Balks mid-river, and we look down into the rapid course
Of the current, though our mount does not budge, still it seems a
    force
Is sweeping his body sideways, swiftly shoving it upstream,
And wherever we cast our glance, every other thing will seem
To be borne along and rushing as ourselves, in the same way.

Next, take a colonnade, with rows in parallel array,
And all of the supporting columns standing at one height,
Yet when you view down its entire length, then in your sight
It seems little by little to taper to the narrowing point
Of a cone, so roof and ground, and left and right, completely
    join,   430
Until, in the vanishing vertex of the cone, they all converge.

To sailors at sea, the sun appears to rise up from the surge,
And to set into the waves again, and drown its light thereunder.
(But since all they behold is sky and ocean, then no wonder –
Don't be too quick to think their *senses* have been deeply
    shaken.)
Again, ships in a harbour seem to landlubbers – mistaken –
To be battling the waves, sterns crippled at the waterline;
For whatever part of the oar lifts up above the dewy brine,

Is straight, and the rudders above the surface, also straight and
          sound,
440 But sunk beneath, they all seem crooked, twisted back around
So that they are bent upwards, and seem practically to ride
Flat on the surface of the water. When winds at eventide
Harry racks of cloud, the sparkling star-signs seem to glide
Against the movement of the clouds, passing high above
Travelling counter to the way they actually move.

Then if you take your hand and press it underneath one eye,
You get a curious sensation; everything you spy
450 Now appears in double vision – double lamps abloom
With flames, twin sets of furniture arranged in every room.
People with two faces and with double bodies loom.
Further, when sleep has tightly bound our limbs in sweet repose,
And the whole frame lies in deepest peace, we nonetheless
          suppose
That we are wide awake, our limbs astir, and that our sight
Beholds the light of day there in the inky black of night.
We think we trade our ceiling for the sky, our cramped room
          yields
To rivers, mountains, sea, we seem to stride across the fields
460 And to hear voices, though the night holds everything
          spellbound
In its grave silence; we seem to speak, but do not make a sound.

And we encounter many other illusions of this ilk –
Amazing sights all striving, as it were, to cheat and bilk
The credulity of the senses; but all for naught, since the lion's
          share
Deceive because of notions that our own minds bring to bear
When they think they 'see' something that the senses did not
          view,
For nothing is more difficult than to distinguish what is true
From false interpretations which the mind applies on cue.

As for the fellow who asserts that 'nothing can be known',
470 He doesn't even know that fact, since he's the first to own

That he knows nothing! I won't debate a person who, instead
Of keeping two feet on the ground, is standing on his head.
Or if I grant he knows that much, I have questions in store:
For since he's never put faith in the sensory world before,
How does he even know what knowing is, or furthermore,
Not-knowing? What forms his notion of the false or of the true,
What evidence has proved the difference between the two?

You'll find the concept of the true is formed and has its root
In the senses, their testimony such that no one can refute.
For there must be a higher court to which you can appeal,   480
That on its own can disprove what is false by what is real.
Besides, on what except the senses can you more rely?
Shall reason, based on the senses' false witness, testify
Against those very senses out of which it's wholly sprung?
For if the senses are untrue, all reasoning is wrong.
Can the ear convict the eye? Or is touch able to bring suit
Against the ear? Can touch against the sense of taste dispute,
Or nostrils confound its argument, or will the eye refute?
No, I think not. Each sense has a function to perform,
A separate jurisdiction. We discern what's soft or warm   490
Or cold, therefore, by one particular sense, and we perceive
The many hues of things and all the qualities that cleave
Closely unto colour by another sense. As well,
It takes the mouth to taste a flavour, but to smell a smell
Requires another sense, another still to pick up sounds,
So one sense can't disprove that which another sense propounds.
Nor can these senses testify against themselves; they must
Be granted at all times an equal measure of our trust.
Thus what they say is true, at any given time, *is* true.

If reason can't unravel the mystery and gives no clue   500
To why what had seemed square to us when viewed from near
    at hand
Looks rounded at a distance, if you cannot understand,
It's better to offer erroneous explanations than let slip
Any aspect of the graspable out of your grip

And that way wreck the fundament of faith, and so lay waste
The whole foundation on which life and life's welfare are
    based.
For not only would all Reason fall apart and come to grief,
But in an instant, *Life itself*, unless you hazard belief
510 In the senses, and back away from the brink of sheer cliffs, and
    avoid
Other places where life and limb are like to be destroyed
And walk the other way. I tell you, all the ranks of word
On word mustered and armed against the senses are absurd!

Lastly, in construction, if the carpenter's rule[11] is bent,
Or if the square is warped on which we base our measurement,
Or if the level anywhere staggers off by even a jot,
All of the structure must be built on crooked lines, the lot
Ramshackle, tumble-down, walls leaning out or in, and all
Out of whack. Now part of the rickety shack is like to fall –
And now part *does* collapse. And all because it was betrayed
By faulty measurements when its foundation was being laid.
520 Likewise your reasoning concerning things is built askew
If founded on sensations that are off from plumb and true.

Now as to how the other senses all perceive, it's plain
Reason won't have a rocky time in trying to explain.
First, take every noise and voice: the reason that we hear
Is that their bodies strike the sense when they crawl in the ear.
Voice and sound are made of matter – admit it, you've no
    choice –
Since they are able to knock against the sense. Besides, the voice
Frequently will grate against the speaker's throat: a shout
Makes the windpipe rough and sore as it is rushing out.
530 Indeed as particles of voice amass in greater force,
And start to pour out through the narrow exit, in due course
The mouth becomes all jammed full, and its 'doorway' is scraped
    hoarse.
Thus words and voices are composed of atoms, it is plain,
Since they possess the power to inflict physical pain.

And you're not wrong about the sheer amounts of mass that
   drain,
Drawn from the very sinews of a man, his might and main,
If he declaims without a break from the first gleam of light
At cockcrow till the shadows gather at the fall of night,
Especially if he holds forth at the top of his lungs. Therefore
It follows that the voice must be constructed out of matter   540
Since a man loses something of his flesh by endless chatter.

And furthermore, whether a sound is smooth or it is gruff
Will depend on its components, whether they are smooth or
   rough.
When the barbaric bugle brays, the elements that come
Off all the hillsides into the ear, booming back the thrum
Of bellowings abuzz with bass, are shaped differently from
Those of the melancholy plaints that in the winding vales
Of Helicon pour melting from the throats of nightingales.[12]

And therefore as the voice is squeezed from deep inside and
   flung
Directly out of the mouth, that nimble artisan, the tongue,   550
Joints it into syllables, while shaping into speech
Is the lips' task. And if the span that each sound has to reach
Across is not too vast, not just the voice, but every word,
Every syllable, can be made out and clearly heard,
Because the sound preserves its shaping and retains its cast.
But if the space a voice must travel is unduly vast,
The words, through such a mass of air, are thrown in disarray,
And while it flies, the voice is jumbled up along the way.
That's how you hear the utterance, while on the other hand,   560
Exactly what the voice is saying, you can't understand,
Since the voice arrives so garbled and tangled up. Again, one
   word
Bellowed from the mouth of the town crier is often heard
By a whole mob of people. So a single voice therefore
Is instantaneously divided into many more,
Since it disperses itself to many an individual ear,
Stamping its own shape and sound on the words, loud and clear.

A portion of the voices fall upon no ear, and these
Pass by and pass away in vain, scattered upon the breeze,[13]
570 And some of the voices, striking solid surfaces, rebound
With an echo, fooling, often, with mere semblance of a sound.

When you have understood this, you'll be able to explain
To yourself or anybody else why rocks make their refrain
In lonely spots, giving back our words in the same array
We call them out to friends who've wandered off and lost their
         way
In the shadowy mountains. I have even known a place reply
As many times as six or seven to a single cry –
So faithfully does hill re-echo unto hill and throw
The voices, guiding their reverberations to and fro.

580 Nymphs and goat-footed satyrs haunt these spots (the locals
         say),
And legend has it there are fauns who dance the night away,
And raise such ruckus with their rowdy revels that they break
The peace and quiet, locals claim, with all the noise they make:
The plucking of the strings, the sweet and melancholy sound
That pours from flutes, as players finger stops. For miles around,
They tell, the farm folk listen when Pan shakes the piney wreath
Of boughs that shades his head (half-goat, half-human)
         underneath,
Ever blowing across the hollow reeds with his sly smile
So the pipes never cease their rustic melodies the while.
They spin out other wild yarns of this nature – by such tricks,
590 Perhaps, hoping to show they aren't so far out in the sticks
That even gods don't visit. Maybe that's why they regale
Us with such curious lore – or other motives might prevail:
The human race is always all-ears for a fairy tale.

As for other mysteries, it should come as no surprise
That voices pass through barriers and lash the ears, while eyes
Cannot see objects clearly through them. Often we overhear
A conversation going on behind closed doors. It's clear

That this is since a sound is able to pass unharmed through bent
Passageways, while images refuse. For they are rent   600
Unless the path is a straight shot, as is the case with glass,
A substance through which any image has the power to pass.

Furthermore, a single voice can scatter on all sides
Since each new voice gives rise to others, once the first divides
And splinters into many – as oftentimes a spark of flame
Divides itself and kindles many other fires the same.
Therefore, even places screened back from our view abound
With voices, and everywhere is seething and aswirl with sound.
But images, once sent forth, tend to travel straight and true:
That's why you cannot see *across* walls you can listen *through*.   610
And yet even those voices that pass through a house's wall
Arrive so dulled and garbled up, that by the time they fall
On our ears, we seem to hear just noise, and not the words at
    all.

And it is so straightforward to explain the sense of taste
On tongue and palate, that any extra effort is a waste.
First of all, in our mouth we taste the flavours when we chew,
Squeezing out the savour from our victuals as we do,
Just as you might squeeze in your fist a sponge that's sopping
    wet
Until it's almost dry. The flavours we press out then get
Dispersed through pores all over the palate, distributed among   620
The tortuous passageways of the more loosely textured tongue.
Then, if the particles of flavour that ooze out are smooth,
They sweetly brush against the tongue, and sweetly touch and
    soothe
All moist and watery places about the tongue. The more they
    tend
To be *prickly*, on the other hand, the more the bodies rend
And sting the senses as they are released. But the delight
Of tasting food is limited to the palate. Once the bite
Is swallowed down the gullet, we do not have a sensation,
As it's distributed throughout the flesh, of delectation.

630 The *kind* of food that nourishes the body doesn't matter,
As long as you can break down what you eat, so it can scatter
All through the limbs, while making sure the stomach's health is
    good.

And now I shall explain why different beasts eat different food,
And why, what's bitter to one animal, or sour to eat,
May, to another creature, taste most wonderfully sweet.
The differences are so vast in this, so varied and askew,
'One man's meat is another's poison' literally holds true.
There is, for instance, a snake that, touched with human spit,
    will die,
Gnawing upon itself until it perishes thereby.
640 Also, take hellebore, a poison guaranteed to wreak
Havoc with us, but which makes goats and quail grow fat and
    sleek.

For you to understand how this phenomenon can be so,
First you should keep in mind what I declared not long ago:
Things are made of many seeds in myriad ways combined.[14]
Consider all the creatures that eat food – just as you find
They differ on the outside, each according to its kind
Having a different outline and proportion to its frame,
Likewise, the shapes of seeds that make them up are not the
    same.
And since the seeds are different, then it is clearly true
650 That the channels and the gaps (which we call 'passageways') all
    through
The flesh, and in the mouth and even the palate, differ too.
And thus some must be narrower, some wider, and some found
To be triangular, and others square, and many, round,
And also an assortment of many-sided shapes abound.
Indeed, the shapes and motions of the particles that make
These passageways determine shapes the passageways must take.
And so the channels vary, according to their differing texture.
Thus what is sweet to one thing may be bitter to another;
One finds it sweet because the bodies that are very smooth
660 Penetrate the portals of the palate, and so soothe.

But the reason it is bitter in another, it clearly follows,
Is that the rough, hooked bodies tear the gullet as it swallows.

Now with these facts in hand it isn't hard to reconcile
All quandaries of this kind. For instance, when excess of bile
Causes a fever, or some other pathogen holds sway,
A man's whole frame is willy-nilly thrown in disarray,
And his component elements are jumbled every way,
And so those particles that fit his senses well before
Are not so well adapted to his senses any more;
Others fit them better now, that are bitter to eat.   670
Indeed, within it, honey has both bitter and the sweet –
Something I've explained to you before, and now repeat.[15]

Now pay attention – I'll address how smell affects the nose.
First, there are clearly many things from which there rolls and
      flows
A complex stream of odours, which we know must scatter
      smells
And spread them hither and thither. But to different animals
Different smells are better suited due to their forms. Thus bees
Are drawn across the air for any distance that you please
By a whiff of honey; vultures by carrion. Hounds in a pack
Lead hunters to their quarry by scenting the cloven-footed track.   680
And people are detected at a distance from their smell
By the snow-white goose, Protector of the Roman Citadel.[16]
And so each type of beast has its own sense of smell, employed
In tracking down its rightful food and helping it avoid
Foul poison. Thus they preserve the generations of their kind.

And yet, of all the odours that greet the nose, although you find
One smell may travel farther than another, even so
No smell can cross the distance that a sound or voice can go,
Not to mention sights that strike the eyes and stimulate   690
Our vision. Smell's a straggler, it meanders, lagging late;
It fades away upon the breeze and meets an early fate.
First, because it emanates with difficulty from deep
Inside an object (and it is clear that smells pour off and seep

From the inner depths of things – since things give off a stronger scent
When they are ground to bits, or burnt up in a fire, or rent).
Then also, you can tell that smell is made of particles
Larger than those of sound, since it cannot go through stone walls,
700 Which, as a rule, sounds and voices pass through. And that's why
It isn't easy, as you'll see, to track where odours lie,
For whiffs of scent grow cold as through the breezes they digress,
Not racing to the nose with news of things hot off the press.
Thus dogs trying to find the scent, stray often from the trace.

As for differing perceptions, it is not only the case
With the sense of smell and taste. Because the aspect and the hue
Of objects are suited differently to different creatures too,
So that some creatures find some sights too harsh for them to view.
710 For instance, take the rooster, who is wont to clap away
The night with flapping wings, and usher in the break of day
With his clarion crow – raging lions cannot stand the sight
Or stand their ground against him, filled at once with thoughts of flight,
Doubtlessly since there are certain atoms that comprise
The body of the cock, which entering the lions' eyes
Dig into the pupils and inflict a sharp pain there,
Such that, for all their fierceness, it is more than they can bear,
While the same atoms do not sting our own pupils at all,
Either because they don't get in, or once inside the ball
720 They're granted easy exit, so that they do not remain
Long enough in any part of the eye to cause it pain.

Now learn about what sets the mind in motion and you'll see,
In only a few verses, how these fancies come to be.
First this: I say that many images of objects stray
To and fro in every quarter and in many a way,

So delicate, they easily stick together when they meet,
As spider webs are wont, or gold to airy thinness beat.
Indeed, the texture of these images is yet more slight
Than that of images which fill the eyes and strike the sight,
Because they penetrate the body through its pores, and quicken   730
The mind's subtle stuff and strike the senses from within.
This is why monsters with their hodgepodge limbs appear to us,
Such as Centaurs and Scyllas, hounds with heads like Cerberus –
And phantoms of the dead, whose bones lie in the Earth's
    embrace –
Because all kinds of images are floating every place.
Some of them spontaneously arise out of thin air,
And some are shed from sundry different objects, and a share
Are formed of combinations of these figures. For it's fair
To say no image of a centaur possibly could derive
(When there is no such thing in Nature) from one that is *live*.   740
But when the images of horse and man do chance to meet,
They easily adhere at once, which is, as I repeat,
Due to their gauzy fabric and the fineness of their texture.
Other hybrids of this sort are formed in the same manner,
And since they travel swiftly, for they are exceedingly light,
As earlier I've demonstrated, any of these slight
Images easily sets the mind in motion with a touch –
The mind's so fine and quick to move that it does not take
    much.
That this occurs just as I say, you easily can find
From the following: since what we see with the eyes and with
    the mind   750
Is similar, then the causes must be similar in kind.
Now therefore, since I have explained before that I descry
A lion, say, by means of likenesses that strike the *eye*,
We can deduce that *minds* are moved in such a manner too,
By likenesses of lions or any other thing they view,
Exactly in the same way that the eyes themselves perceive
Except the images minds see are flimsier in weave.

How else could it be that, when in sleep the limbs go slack,
The mind itself is active, unless the same films that attack

The mind when we're awake, besiege our sleep – such that it
    seems
760 That we behold him in the flesh, when we see in our dreams
A man fled from this life, and in the power of death and dust.
But there's a natural explanation: since all sensation must
Be lying checked and listless in the limbs, it can't appeal
Against the witness of the False by evidence of the Real.
Besides, the Memory lying lax in sleep cannot avow
The fellow's long been in the grip of grave and death by now,
Whom the mind itself believes that it beholds alive and well.

Further regarding likenesses, it is no miracle
An image moves its arms and other limbs about in time
770 (For often in our sleep an image seems to pantomime)
Since when one image vanishes, the next that rises shows
Another stance, so that the first one seems to shift its pose.[17]
Of course you must consider that this happens very fast,
Their swiftness is so great, and the supply of things so vast,
And the store of particles which objects give off so immense
It can supply a sight for every moment that we sense.

This subject raises many questions; many things remain
For us to clarify, if we would make the matter plain –
For one thing, when the mind desires to picture something, why
780 Is it able to see it right away? Do images stand by
Till they are wanted by us, so the instant that we please,
The image that we wish occurs to us, be it the seas,
Or earth, or even the sky above that we are longing for?
Parades, throngs of people, feasting, skirmishes of war –
Does Nature fashion and supply such pictures on demand?
(Especially since there are other people near at hand
Imagining sundry things of a completely different sort!)

Then what about in dreams when we see effigies disport,
790 Moving supple limbs in rhythm, as the nimble dancers
Sway their tender arms in turn, and their light footwork answers
With corresponding steps before our eyes? We should conclude,
I take it, that these roving images have been imbued

With art, rehearsed to frolic in their nightly interlude!
Or will the explanation be the following, instead:
In any perceived instant, say, in the time one word is said,
There exist many further divisions of Time lurking still
(The mind grasps this), and that is why at any time you will,
All images are ready to hand wherever you are – so fast
Is the speed at which they fly, and the supply of things so vast.
Thus, when one image passes away, the next that rises shows   800
A different stance, and so the first one seems to change its pose.
And since they are so fine, the mind cannot perceive them clear
Unless it makes an effort. Thus, all of them disappear
Except the images the mind has geared itself to see.
Yes, the mind prepares itself, and looks expectantly
For the image that comes next. And therefore, come the image
     does.
Don't you see that even the eyes, when they begin to focus
On something small and delicate, prepare themselves to strain,
And if they did not do so, we could not discern it plain?   810
Yet even regarding obvious objects, you will find it's true
That unless we train attention on them, they recede from view,
Fading into the background. So why be surprised to find
That every other thing escapes the notice of the mind
Save what the mind is focused on? And this creates illusions –
We draw from evidence of flimsy clues sweeping conclusions,
And get ourselves entangled in the web of our delusions.

Sometimes it also happens the next image to arise
Differs in kind from the preceding one: before our eyes
What was a woman becomes a man, and faces change their guise   820
Or age transforms, and yet we are not startled at the change –
Sleep and Forgetfulness ensure we do not find it strange.

A mistake I strongly urge you to avoid for all you're worth,
An error in this matter you should give the widest berth:
Namely, don't imagine that the bright lights of our eyes
Were purpose-made so we could look ahead, or that our thighs
And calves were hinged together at the joints and set on feet
So we could walk with lengthy stride, or that forearms fit neat

To brawny upper arms, and are equipped on right and left
830 With helping hands, solely that we be dexterous and deft
At undertaking all the things we need to do to live.
This rationale, and all the others like it people give,
Jumbles effect and cause, and puts the cart before the horse –
For nothing is born just so that we can use it – in due course,
That which is born creates its *own use*. Before the light
Of eyes arose, there was no such thing as a sense of sight.
Before the tongue was fashioned, there were no words to recite.
But rather, the genesis of the tongue by far pre-dates the word,
And ears came into being long before a noise was heard.
840 In short, the organs and the limbs existed, I surmise,
Before there was any use for them.[18] Thus they did not arise
For the purpose of performing certain functions. No, instead,
In combat, fighting hand to hand had already existed,
And ripping of the limbs, and fouling of the flesh with gore,
Long before the launch of gleaming javelins in war.
And human nature urged men to avoid coming to harm
Before a soldier learned to wield a shield on his left arm.
And surely, to succumb to sleep to rest a weary head
Antedates by far the cushioned mattress of a bed.
850 Before there were cups to drink from, there existed thirst to
     slake.
That these inventions were devised for usefulness's sake,
From the experience of living, we can easily suppose.
But concerning all those other things, however, that arose
Before an inkling of their use, it's not the case at all –
We see, particularly, that the limbs and senses fall
Into this category. This is why I say again,
You can't believe they were created to fulfil a function.

Another fact that's no surprise – the body of each beast
By instinct seeks the nourishment upon which it must feast.
860 Indeed, I've shown that many atoms flow off and are cast
From objects in a variety of ways, and yet the vast
Majority are lost from animals. For since they keep
In motion, myriads of atoms are squeezed out from deep
Inside their bodies and in the form of perspiration seep

Away, and many through the mouth are lost when they exhale,
Panting and tired. Thus losing density, the frame grows frail,
Its whole structure undermined. Pain follows on that score,
And that's the reason nourishment is taken in, to shore
Up the subsiding limbs, and to renew the body's might,
Distributed through the veins, to fill its gaping appetite.
Liquid also trickles to all parts that beg for water,   870
And the many particles of heat that are clumped up together
And kindle a burning in our stomach, all these bodies scatter
When fluid, flooding in, extinguishes them like a flame,
So that the parched heat can no longer burn within our frame.
This is how the panting thirst is sluiced away and spilled
From our flesh, how the famished yearning's satisfied and filled.

Now I'll explain how we can walk when we decide to go,
And how it's in our power to ply our members to and fro,
And by what process we are habitually able to convey
The body's heavy freight. So listen close to what I say: 880
I tell you, first, an *image* of walking comes into the mind,
And strikes it, as I've said before. Desire follows. You'll find
That no one starts to undertake a single act until
The mind has looked ahead and has decided that it will.
(Whatever thing the mind foresees, there is an image of.)
And therefore when the mind rouses itself and wants to move
And walk, it stirs the whole force of the spirit scattered through
The limbs, all in a trice. And it's an easy thing to do,
Seeing that spirit and the mind so closely intermesh.
And spirit in turn prods the body, so all the mass of flesh   890
Is pushed ahead and moved little by little. Then, besides,
The body becomes more porous, and the air thus comes inside
(Naturally enough – it is so mobile), through the wide-
Opening passageways, abundantly, dispersing all
Throughout the body, down to every part, however small.
Therefore the body's borne along two different ways, by twinned
Forces, as a ship is borne along by sail and wind.

And on this subject, it should not elicit your surprise
That particles so small can turn a body of this size

900 And wheel the whole cargo of our weight. For though the stuff
     Of wind is fine and delicate, yet still it is enough
     To shove along the huge bulk of a hulking ship. Indeed,
     It takes but a single hand to steer a ship, whatever speed:
     A single rudder turns it – any direction, it must heed.
     And machines, by means of rollers and of pulleys, can transport
     Or hoist aloft many a hefty load with little effort.

     Now then, regarding sleep and how it floods the limbs with rest,
     Loosens the mind from troubles and unknots cares from the
         breast,
     That is what I shall explain in verses short but sweet –
     Just as swan song is sweeter, even though it may be fleet,
910 Than the endless squawking of a flock of cranes in the southern
         sky,
     Lend me your sharp ears, and a keen mind – lest you deny
     That what I say is possible, and so you don't depart
     Hardened against the truth of what I tell you in your heart,
     Though if you cannot see the truth, you've just yourself to
         blame.

     First, sleep occurs when the spirit's force is scattered through the
         frame –
     So that some parts are cast abroad, and some parts of it keep
     Within, crowded together more, retreating way down deep –
     And only then do the limbs relax and slacken. For it's clear
920 This consciousness we have must come within the spirit's sphere,
     So that when sleep disrupts this consciousness, we must decide
     The spirit has been scrambled up and been expelled outside –
     Not all of it, of course, for then the body would lie still,
     Steeped deeply in the freeze that is Death's everlasting chill.
     For if no portion of the soul remained, hid in the members
     (As fire may lurk, buried under a mound of ash and embers),
     Then how could consciousness become rekindled through the
         frame
     All of a sudden, just as, from a banked blaze, rises flame?

But what it is that brings about this new condition, how
The spirit is disorganized, and limbs go limp, I now   930
Shall explain – and you, make sure my words are not poured
    forth in vain
To the four winds. For first, the outer body must sustain,
As it comes into contact with surrounding air that flows
All round it, a continuous rain of buffetings and blows.
(And it's on this account that nearly everything, you'll mark,
Is covered by a kind of skin, or shell, or peel, or bark.)
And at the same time air is lashing our body from within
When we breathe, drawing air inside and blowing it out again.
And therefore, since the body's flogged inside and out, and
    more
Blows penetrate through every minute opening and pore   940
To the body's fundamental particles and elements,
Bit by bit, our bodies are falling to pieces, in a sense,
As atoms of mind and flesh shift from the places they abide,
And part of the spirit is cast forth, and part withdraws to hide
Deep within, and what's left over, scattered far and wide
Through the frame, cannot join up or carry out reciprocal
    motion –
Nature blocks the pathways and cuts off communication.
Thus, its motions disrupted, consciousness recedes down deep
Inside, and since there now is nothing, as it were, to keep   950
The frame propped up, the body slumps, and all the limbs go
    slack,
The arms and eyelids sink, and as you lie down, often the back
Of the knees give way beneath you, all the strength of them
    unstrung.

Next, eating leads to drowsiness, since food, dispersed among
The veins, has the same effect as air. You fall in deepest
    slumber
When you're exhausted, or full, because it's then the largest
    number
Of particles, hammered with heavy toil, are thrown into
    confusion.
Part of the spirit contracts deep inside for the same reason,

960 And a larger amount of spirit is cast from the flesh, outside,
And its particles within are scattered farther apart and wide.

And whatever interest fascinates us, whatever thing we make
Our business, what occupies the mind when we're awake,
Whatever we're most focused on, it is that thing, it seems,
That we are likeliest to meet with also in our dreams:
Advocates keep arguing cases, and have claims to settle,
High commanders take the field and lead troops into battle,
Sailors keep on duelling it out with the winds, their sworn foe,
And even I strive at my work – ever seeking to know
970 The Nature of Things, and scratching down my findings as I go
In our mother tongue. And thus the other arts and callings keep
Their hold upon men's minds with idle fancies while they sleep.

And then we often see when people have been hard at play,
Devoting themselves only to spectacle, day after day,
Even when they stop, and leave those sights and sounds behind,
There are still corridors that stay propped open in the mind
Through which the images of these can enter – so for days
Afterwards, the images appear before their gaze,
980 And even when they're wide awake, they see the dancers swaying
Their nimble limbs, and hear snatches of liquid music playing
From the singing strings of lyres, and they behold the stage,
        along
With all its fluttering pomp and glory, and the familiar throng.

That's how much we're affected by those interests we pursue –
Our habits and our hobbies and the business that we do.
And not just men: the same is true for creatures of all breeds.
In fact, you'll see, though lying down, legs sprawled, that
        spirited steeds
Are sweating and are blowing hard as if they ran a race,
Galloping down home-stretch with all their might to win first
        place,
990 Or as if out of opened starting gates, they strained to leap.
And frequently the hounds of hunters, although fast asleep,
Suddenly start twitching their paws and whimpering, and keep

Snuffing the air with nose a-quiver, as if hot on the trace
Of their wild quarry; and if you wake them up, they often chase
The spectral images of stags they see in headlong flight,
Until they've shaken the spell, and come back to themselves
    aright.
Then take the affectionate brood of house-bred puppies – quick
    to rise
To their feet, and shake themselves, as if in curious surprise
At a stranger's form and figure they behold before their eyes.[19]   1000
(The more ferocious that its species is, the more it seems
A creature has to be ferocious even in its dreams.)
Birds of many a feather, without warning in the night,
Disturb the groves of gods with flapping wings as they take flight
If in their gentle slumber they catch sight of hawks that soar
And swoop down after them with battle cries declaring war.   1010

And then, the minds of men, which can accomplish any number
Of great deeds by great motions, often repeat them in their
    slumber.
Kings conquer, and in turn are captured, join the battle's fray,
And whimper as if their throats were slit, although they never
    stray
From bed. Many in throes of mortal combat groan their pangs,
And as if being mauled by panther's or cruel lion's fangs
Fill their chamber with bloodcurdling screams. Many in sleep
Babble on and spill important secrets they should keep,
And oftentimes confess their guilt. Many meet their death.
Many feel their body hurtling headlong towards the earth   1020
From dizzy mountain heights, and crazed with fear, can scarcely
    waken
And come back to their senses, half out of their wits, they are so
    shaken
By hammering heart-beats. A thirsty man will often kneel hard
    by
A stream or cooling spring and try to drink the river dry.
A little child when lying fast asleep will often think
He's lifting up his robes beside a chamber-pot or sink

And pours forth all his body's filtered liquid – and so wets
The bed, and all the splendour of Babylonian coverlets.
1030 For those in adolescence's riptide, when Manhood has made
Seed in their limbs for the first time – then images invade,
Images of some random body or other – bringing news
Of a lovely face and radiant complexion's rosy hues.
This irritates and goads the organs, swollen hard with seed –
Such that frequently, as if he'd really done the deed,
A youth floods forth a gush of semen so he stains the sheet.[20]

This seed is agitated in us, as I say again,
When onset of adulthood gives our limbs their might and main.
For different things are stimulated each a different way –
1040 To stir up human seed from man requires human sway.
This seed, as soon as it is shaken loose out of its seat,
Drains from the limbs, and from the frame at large makes its
        retreat,
And collecting in certain areas of the groin, it concentrates,
And suddenly arousing the genitalia, stimulates
And swells them up with seed. There follows a desire that yearns
To eject seed towards the one for whom the lust, disastrous,
        burns.
The body seeks what struck the mind with love and caused it
        hurt.
For as a rule, men fall towards the wound, and blood will spurt
1050 Along the same trajectory from which we took the blow;
And if the fighting's hand to hand, a red gush splatters the foe.
Therefore if someone's injured by the point of Venus' dart,
Whether a boy with girlish limbs has launched it at the heart,
Or a woman shooting the arrows of love from her entire frame –
He seeks the source and longs to mingle with it all the same,
To cast his fluid to another body – appetite
Wordlessly anticipating pleasure and delight.

This is what our Venus is. It is to this we impart
The name of Love – and it was this first dripped into the heart
1060 Venus' honey – and Icy Care has followed in its train.
For though your love be absent, still the images remain,

The darling name of the beloved ringing in your ear.
It's best to flee away from images, and to steer clear
From the fodder that love feeds upon – it's better to direct
Your attention somewhere else, and spend the fluids that collect
On any body – rather than retain them and remain
Fixed ever on one love, laying up stores of certain pain.
Feeding the sore will only make it fester and grow strong –
For madness and misery grow graver as time goes along
Unless you reopen the first wound with new cuts while it's still
    fresh   1070
And with the Venuses of Easy Virtue cure the flesh,
Or turn your mind to something else, to some other pursuit.

Nor does the man who gives the slip to Love lack for its fruit;
Rather, he enjoys it without penalty or pain:
Pleasure's unalloyed not for the lovesick, but the *sane*.
Even at the very moment of having, the raging tide
Of desire tosses lovers this way and that. They can't decide
What to enjoy first with hand or eye – so closely pressing
What they long for, that they hurt the flesh by their possessing,
Often sinking teeth in lips, and crushing as they kiss,   1080
Since what the lovers feel is not some pure and simple bliss –
Rather, there are stings that lurk beneath it, pains that shoot,
Goading them to hurt the thing that's made madness take root,
Whatever it may be.

                  But Venus makes their suffering light
In the midst of love, and Pleasure, mingled in, curbs back the
    bite.
For herein lies the hope: they think that they can quench the fire
By means of the same body that ignited their desire,
Something Nature contradicts with all her might. For love
Is unique: the more we have of it, the more it's not enough,
And the more calamitous desire sets the heart aflame.   1090
As for food and drink, they are absorbed deep in the frame,
And since they can fill up the areas that are set aside,
Craving for water or for bread is easily satisfied.

But of a human face's bloom and beauty, what comes in
For the body to enjoy? Just images, flimsy and thin,
And the wind often snatches even this scrap of hope away.

As in a dream, when a man drinks, trying to allay
His thirst, but gets no real liquid to douse his body's fire,
And struggles pointlessly after mere images of water,
1100 And though he gulps and gulps from a gushing stream, his
        throat is dry,
So Venus teases with images – lovers can't satisfy
The flesh however they devour each other with the eye,
Nor with hungry hands roving the body can they reap
Anything from the supple limbs that they can take and keep.
Lastly, when their limbs are tangled, and they pluck youth's
        bloom,
And bodies have a foretaste of the pleasures that now loom,
And Venus is about to sow the woman's field with seed,
They grasp each other and mix the moisture of their mouths in
        greed,
And panting heavily, press teeth in lips, but all in vain –
1110 There's nothing of the other they can rub off and retain.
Nor can one body wholly enter the other and pass away –
For it seems sometimes that this is what they struggle to essay,
Such do they clasp in the chains of Venus, greedily and tight,
While limbs go limp, melted with the heat of their delight.
At last, when loins erupt forth from the gathering desire,
They are allowed a brief reprieve from passion's raging fire.
But then the fever starts again, madness must soon return,
When yet again they seek to have the thing for which they yearn.
They can discover no device to conquer their disease –
1120 But waste away from wounds unseen, amidst uncertainties.

Add this – lovers fritter away their strength, worn out in thrall.
This also – one lives ever at the other's beck and call.
They grow slack in their duties. Good name stumbles and
        malingers.
Wealth, turned to Babylonian perfumes, slips through the
        fingers.

But you can bet that *she*'s well heeled, in shoes from Sicyon,
And those are genuine emeralds, the rocks that she's got on.
The wine-dark sheets, from rough and constant use upon the
    bed
And drinking up the sweat of Venus, are worn down to the
    thread.
The father's hard-earned fortune turns to tiaras for her hair,
Alindan silks, diaphanous gowns from Cos[21] for her to wear.   1130
He shells out for fantastic feasts with all the trimmings – fine
Linens, music, perfume, garlands, wreaths, free-flowing wine –
But in vain – since in the very fountain of delights, there rises
Something of bitterness that chokes even among the roses.
Perhaps it's that remorse, gnawing at the conscience, taunts
The lover he's thrown his life away in sloth, among low
    haunts;
Or else his darling wings a two-edged word at him, a dart
That smoulders like a fire, and rankles in the love-struck heart;
Or else he thinks her roving eye too freely wanders after
Another, and imagines in her face a trace of laughter.   1140

And these are just the problems of a love that's going *well*!
Imagine a love that's crossed and doesn't have a chance in
    hell –
Even with your eyes shut, you can grasp that the amount
Of troubles in unhappy love are more than you could count.
Best to keep eyes open, as I've said – don't take the bait.
It's easier to avoid the toils of love than extricate
Yourself once you are caught fast in the nets and to break free
From the strong knots of Venus. Yet you're still able to flee
The danger, even if you're tangled up, snared in the gin,
So long as you don't stand in your own way, and don't begin   1150
To overlook all shortcomings in body and in mind
Of the woman you lust after. For desire makes men blind –
And generally they overlook their girlfriends' faults, and bless
These women with fine qualities they don't in fact possess.
That's how it comes that we see girls – malformed in many
    ways,
And hideous – are petted darlings, objects of high praise.

Indeed, one lover often urges another he would mock:
'Venus has it out for you – your love's a laughing stock.'
(Poor fool – that *his* delusion's worse would come as quite a
    shock!)

1160 The black girl is *brown sugar*. A slob that doesn't bathe or clean
Is a *Natural Beauty*; *Athena* if her eyes are greyish-green.
A stringy bean-pole's a *gazelle*. A midget is a *sprite*,
Cute as a button. She's a *knockout* if she's giant's height.
The speech-impaired has a *charming lithp*; if she can't talk at all
She's *shy*. The sharp-tongued shrew is *spunky*, a little *fireball*.
If she's too skin-and-bones to live, she's a *slip of a girl*, if she
Is sickly, she's just *delicate*, though half dead from TB.
Obese, with massive breasts? – a *goddess* of fertility!
Snub-nosed is *pert*, fat lips are *pouts* begging to be kissed –
1170 And other delusions of this kind too numerous to list.
Yet even if her face has every beauty you could name,
And she pours out the power of Venus from her entire frame,
The truth is, there are other fish in the sea. The truth is, too,
We've lived without her up to now. She does – we know it's
    true –
Exactly the same things as all the ugly women do,
And fumigates herself, poor girl, to cover the stench after,
While her maids steer clear of her and try to hide their
    laughter.[22]
But the lover, locked out, weeps, and strews the stoop with
    wreaths in bloom,
And anoints the haughty doorposts with sweet-marjoram
    perfume,
And presses his lips to the door, the fool[23] – when if he were let
1180   in,
One whiff and he would seek a good excuse to leave again!
His long-rehearsed heartfelt lament would then come crashing
    down,
Right then and there he'd curse himself for being such a clown,
And for granting her perfection that no mere mortal attains.
Our Venuses are on to this – that's why they take great pains

To hide the backstage business of life, keeping unaware
Those whom they wish to hold bound fast, caught in desire's
    snare.
But all in vain, because your mind can drag everything out
Into the light, and find what all the tittering is about –
Yet if she is good-natured, never spiteful, it's only fair   1190
To make allowances for foibles that all humans share.

Nor is a woman always faking passion when she heaves
With sighs, as she embraces a man, and to his body cleaves
With hers, and sucks his lips, and with her kisses wets his face.
For often she does it from the heart, and seeks when they
    embrace
Mutual joys, and spurs him to run the full length of Love's
    race.
Take birds, wild beasts, cows, ewes and mares – for what other
    reason
Would they submit to the male, unless, when they come into
    season,
Their nature overflows, being in heat, and so they thrust
Against the member of the covering male with joyous lust.   1200
And haven't you seen those whom mutual pleasure binds
    together,
How they are tortured in the bondage of their common
    tether?
For often at the crossroads we see mating curs get stuck:
They tug apart, wanting to go their separate ways – no luck,
Since all the while the sturdy fetters of Venus bind them tight –
They'd never do it, unless they felt a mutual delight
Which has the power to lure them to the trap, and to ensnare
Them fast. The pleasure, I say again, is something that *both*
    share.

And as the seeds mix, when the woman happens to prevail,
Her force suddenly conquering the power of the male,   1210
The children resemble the mother, from maternal seed; in turn,
When male seed dominates, it is the father you discern.

And when you see both parents' features mixed up with each
　　other,
The children come from flesh and blood of both father and
　　mother
When the seeds, goaded by Venus, were aroused throughout the
　　frame,
And met as the two panted together, both burning with one
　　flame,
Neither seed conquering or conquered, both counting the same.

Sometimes children take after their grandparents instead,
Or great-grandparents, bringing back the features of the dead.
This is since parents carry elemental seeds inside –
1220　Many and various, mingled many ways – their bodies hide
Seeds that are handed, parent to child, all down the family tree.[24]
Venus draws features from these out of her shifting lottery –
Bringing back an ancestor's look or voice or hair. Indeed
These characteristics are just as much the result of certain seed
As are our faces, limbs and bodies. Females can arise
From the paternal seed, just as the male offspring, likewise,
Can be created from the mother's flesh. For to comprise
A child requires a doubled seed – from father and from mother.
1230　And if the child resembles one more closely than the other,
That parent gave the greater share – which you can plainly see
Whichever gender – male or female – that the child may be.

And it is not the supernatural realm that is to blame
When a man's infertile – so that he will never hear the name
Of 'father' from precious progeny, and childless, he grows old –
Though this is the belief that the majority of men hold,
As gloomily they stain altars with rivers of blood and burn
Sacrifices – that the abundance of their seed return
And make their wives heavy with child. But they pester gods in
　　vain.
1240　Some are sterile because their semen's too viscous – then again
Some are sterile because it is too runny and too thin:

The thin because it can't attach to the womb – it will not stick,
But suddenly runs, sloughed off in a miscarriage; while the
  thick,
Too concentrate, either doesn't go forth with sufficient force,
Or cannot enter the area it should, being too coarse,
Or once there, cannot mingle with the woman's seed with ease.
Compatibility in Venus differs by wild degrees –
Some men get certain women pregnant more easily, and then,
Some women more easily conceive and bear from certain men.   1250
And many women barren in former marriages can get
Husbands afterwards by whom they can have babies yet,
And be blessed with beloved offspring. Often, also, men
Whose wives, though fruitful in previous wedlock, in their
  homes are barren,
Have later found themselves a mate whose nature fit with
  theirs,
Thus they could fortify the days of their old age with heirs.
The key is that the seeds are able to mingle and combine
In such a way that they conceive – the thick seed with the fine,
The watery with the thick. Even the victuals you partake   1260
Of matter – some cause the seed to thicken in the flesh. Some
  make
The seed grow watery, on the other hand, and waste away.

The pleasurable deed itself will also have a vital say –
It's important *how* you do it. People generally believe
That wives more readily in the manner of wild beasts conceive,
For it's in this position that the seed can occupy
The right place, with a lowered breast, and with the loins raised
  high.
Wanton wiggling's of *no use* for wives – no, not one bit –
For a woman prevents pregnancy this way, resisting it,
When she grinds her buttocks against the man's member as it
  thrusts,   1270
Gyrating, her whole body turned to jelly with her lust.
By doing this, she turns the furrow away from the straight and
  true
Path of the ploughshare, and the seed falls by the wayside too.

Whores thus have their own reasons for wriggling – so that they
   can
Spend less time pregnant, and to make it better for the man.
Clearly, though, our wives can have no use for such an art.

Nor is the power of a god to blame, nor Venus's dart,
When, from time to time, a plain girl steals somebody's heart.
1280 Sometimes it is the woman herself who has achieved this feat;
By winning ways, and keeping her dress and person clean and
   neat,
She makes it easy to learn to live with her. And yet above
All this, it is familiarity that leads to love.
For anything that's hammered by a blow, day after day,
However softly struck, at length is conquered and gives way.
Haven't you seen how drops of water falling, on their own
Have the power, over time, to wear their way through stone?

BOOK V

COSMOS AND CIVILIZATION

Who can build a fitting song, who has the strength of heart
To match the Majesty of Things and these truths in his art?
Or who has such a way with words his praise can match the
  worth
Of him[1] who sought these revelations and who brought to birth
Out of his own intellect such gifts of wondrous good
And then bequeathed them to us? None I think, of flesh and
  blood;
For Memmius, to speak in the exalted tones we need
For the Majesty of Things – he was a god, a god indeed,
Who first discovered this way of living life that we now call
Philosophy – for having found Life tossing in a squall,   10
He used his science in the dark and murky storm to steer
Into calm waters[2] and safe harbour, where the sky was clear.

Take for example those ancient discoveries we call divine:
It's said that Ceres taught men to grow grain, and as for wine,
That Bacchus introduced the culture of the clustered vine.
And yet life can be lived without these discoveries, for they say
That there are tribes[3] that live without them to this very day.
But life was *not* worth living till the heart was purified.
Thus he deserves his godhead more, whose Word now far and
  wide
Being broadcast among the mighty nations, sweetly soothes   20
Troubled spirits with life's consolation, his great truths.

But if you think the deeds of Hercules[4] compare somehow,
You stray from truth and common sense. For what harm could
    come now
To us from the gaping jaws of the Nemean lion? And what
    more
Have we to fear now from that bristly brute, the Arcadian boar?
What danger does the Cretan bull now pose? And who now
    shakes
At Lerna's scourge, the Hydra picketed with poison snakes?
What threat to us is threefold Geryon with his triple breast?
Or those dire fowl that out by Lake Stymphalus made their
    nest?
30 And the fire-breathing mares of Diomedes – would we face
Any peril from *them*, stabled far away in Thrace?
Or take the guardian of the Hesperides' apples of bright gold –
The serpent fierce with searing stare whose massive coils enfold
The tree's trunk – what harm could it do now, living beside
The far Atlantic shore washed by the Ocean's pitiless tide –
A place we never approach, where even savages fear to tread?
And what of all those other slaughtered monsters that lie dead –
If they had not been slain, what damage could they do alive?
None at all, I tell you, for even now on earth there thrive
Many species of savage brutes, fierce animals in droves,
40 And chattering Terror teems in mountain lairs, deep woods,
    dark groves –
The kinds of places as a rule we never need go near.

And yet what dangers threaten if the mind is not washed clear,
What battles we unwillingly invite into the heart!
How biting are desire's cares that worry man apart,
How menacing the fears! And then consider Pride and Wrath
And Lust – and the catastrophes which are their aftermath –
50 And Gluttony and Sloth.[5] And he who's conquered all these,
    then,
And banished them from the mind – not by the sword, but by
    the pen –
Shouldn't he be numbered with the gods and not with men? –

Especially because of the holy wisdom he would preach
Concerning the deathless gods themselves, and since, when he
    would teach,
He unfolded the whole Nature of the Universe in his speech.

I follow in his footsteps, while I track his trail of thought,
Teaching in these verses by what law all things were wrought,
And which they must obey, since they have not the strength to
    break
The unbending commandments of Time. Primarily, for instance,
    take
The mind, material by nature, which has a birth, we've found,   60
And which can't last a lengthy span of time intact and sound,
And how it's only *images* that cheat the mind in dreams
When we behold someone who's fled this life, or so it seems.

To resume: I've reached the juncture of my argument where I
Must demonstrate the world too has a 'body', and must die,
Even as it had a birth. And I must show likewise
How the assembly of matter laid the groundwork for the skies,
The earth, the sea, the constellations, the sun, the lunar ball;
What creatures sprang from earth, what beasts were never born
    at all,   70
And how it was men came to use the varied sounds of speech
By taking up the names of things, when talking each to each,
And in what way that awe of the gods slipped into the heart
Which on the globe of earth keeps some things sacred and apart:
The altars, shrines, lakes, groves, gods' graven images in art.
I shall explain as well how Nature, at the rudder, steers
The movements of the moon, the course on which the sun
    careers,
Lest we perhaps suppose between the earth and sky they go
Traversing their yearly pilgrimage, because they will it so,
Thoughtfully assisting corn to ripen, beasts to teem –   80
Or that they orbit due to some divinely ordered scheme.
For even those who are well versed in how gods live without
Care and strife, if they start wondering how things come about,

Especially those events that they see happening overhead
In heaven's regions, backslide into superstitious dread.
They take up harsh taskmasters, whom the wretches think are
    All-
Powerful, because they do not know what's possible
And what is not, namely, that each thing's power has to keep
90 Within certain limits, and has its boundary stone set deep.

But no more stalling you with promises – behold the sea,
The earth, the sky above us, Memmius, the trinity
Of Nature, and her triple body, three forms that display
Such different aspects, three such textures. Yet a single day
Will deliver all of these to doom – and all the world's vast
Mass and machinery, held up age after age, at last
Will come collapsing down. I know this must seem strange and
    new,
The ruin that the heavens and the earth are headed to,
I know how hard my task will be to win you to this view;
100 Such is the case when one brings alien concepts to men's ears,
Something one cannot set before their eyes so it appears,
Nor place it in their hands (for it's by touch belief is brought
Straight down faith's highroad to the heart, into the realms of
    thought).
But I'll speak anyway – events themselves perhaps will prove
My words true – any moment now, you might feel the earth
    move
Beneath our feet, and everything shake to shivers in violent
    quakes.
May Fortune steer this course far, far from us for all our sakes –
May Reason convince us, not the event itself, as cracked
    asunder,
The entire world collapses, crashing with ear-shattering thunder.

110 Before I launch in, though, upon this subject, saying sooth
With more of inspiration and a greater grasp of truth
Than the Pythia who prophesies inhaling fumes of bay
From Apollo's tripod, I'll show you many things that will allay

Your fears, set forth in words of wisdom, so you don't surmise,
Hagridden by Religion, that the earth, the sun, the skies,
The sea, the stars, the moon, are bodies holy and sublime,
And made of godly stuff that must endure throughout all time,
And so you don't believe it should be treated as a crime
As heinous as the one the Giants were guilty of, to bring
The ramparts of the world tumbling down with reasoning;
Or to wish to snuff the sun out, brightly shining in the sky,   120
By smirching such immortal things with rumours they will die.
These objects are so far removed from divinity, so at odds,
And are so little deserving to be numbered with the gods,
That, rather, they supply the paradigms for us instead
Of what lacks quickening movements and sensations and is
        dead.

In actuality it is not possible to find
In every single body an intelligence and mind –
Just as the trees can't take root in the sky, nor can clouds scud
Through salty seas, nor fishes school out in the fields, nor blood
Ooze through bark, nor sap arise in stones. Rather, all things
        know   130
A fixed, allotted habitat where they can thrive and grow.
Neither is mind able to arise, sans flesh, alone,
Nor thrive at far remove from blood and tissue, on its own,
And even if it could, wouldn't the mind's power instead
More plausibly be seated in the shoulders, or the head,
Or even the bottoms of the feet – indeed be born and nestle
In *any* part – so it stays in the same person and vessel?
But since within our bodies we see there is set aside
A certain area where spirit and the mind abide,
And where they grow, all the more reason that we must not
        claim   140
That they can thrive outside the flesh, without a living frame,
In rotting clumps of dirt, or conflagrations of the sun,
Or water, or the towering reaches of heaven – therefore none
Of these bodies is endowed with the divine sensation, seeing
None of these can be animated into a living being.

Another belief you should avoid is that the holy thrones
Of the gods exist here anywhere within our world's zones;[6]
For the essence of the gods is one so subtle and refined,
So far removed from our senses, it is scarce grasped by the mind.
And since it escapes the feel of hands, and slips beyond our
150   touch,
It cannot impact anything else we feel – since inasmuch
As something can't be *felt* by touch, it cannot *touch*. And so
The dwellings of the gods are not like our abodes below;
Rather, they must be, as their bodies are, of flimsy stuff –
Something I'll demonstrate to you in detail soon enough.[7]

And further, Memmius, to say it was the immortals' plan
To create the splendid nature of the universe for *Man*,
And that, therefore, it's right we praise it as the gods' sublime
160 Masterpiece, and deem it deathless, lasting throughout all time,
Or think that something chartered for Man's benefit forever
By the gods' ancient decrees is therefore something that can
    never
By law be shaken from its roots by any force whatever,
Nor rattled and turned topsy-turvy by an argument –
Or any other claptrap of this sort one may invent –
Is ludicrous! For what could such immortal, blessed powers
Stand to gain from any lavish gratitude of *ours*
That they should lift a finger for *us*? Or what new thing
    transpire
That could tempt beings so serene *before* to now desire
To change the life they had already? For, if truth be told,
170 Who revels in newfangledness is fed up with the old.
But what could spark a hankering in someone for the new
Who's never encountered any woe in all time hitherto,
Who lives a life of bliss? If we had never been created –
What would have been the harm to us? Am I to be persuaded
Our life lay cringing in a mire of misery and dark
Until the dawning of Creation lit us with its spark?
For anyone who has been born desires to hold on tight
To life – at least as long as he's detained by sweet delight –

But he who's never tasted lust for life, nor was among
The roll-call of the living, then how can it do him wrong   180
That he was never made at all?

                               Again, where could gods find
A model for creating things – what planted in their mind
The notion of mankind, so they knew what they undertook
To make, and they could picture in their hearts how it should
    look?
And how did they discover atoms, how were their powers found,
And what could be accomplished just by shifting them around,
Unless Nature herself supplied them with the paradigm
For creating things? Since up to now, throughout unending
    Time,
So many different atoms struck by every kind of blow,
Or carried along under their own weight, were wont to go
And meet in every way, and try out every combination   190
To find out what could be created by their congregation,
That they should fall into such patterns should be no surprise,
Nor that they came into the very motions that comprise
The Sum of Things that now renews itself before our eyes.

But even were I in the dark about what atoms are,
I would, from the very workings of the skies, venture this far –
And muster up a host of other evidence to make
My point – the universe was not created for our sake
By powers divine, since as it stands it is so deeply flawed.
First – of the regions that the sky tents over in its broad   200
Sweep – the lion's share is held by mountains and woods, the
    dens
Of savage beasts, and part by rocky crags and desolate fens,
And the sea holding the dry land's shores apart at such great
    length.
In fact, of this land, almost two-thirds, due to the sun's strength,
Or else the ceaseless fall of frost, are pilfered from mankind.
And even what little arable land there *is* that's left behind,
Nature herself would choke with thorns, unless by toil and strife
Mankind fought back, groaning over the mattock for dear life,

And underneath the pressure of the plough, cut open the earth.
If with our labouring we did not urge the land to birth
210 By turning over fertile clumps of clay with the plough's share,
And breaking the land, crops could not rise into the liquid air
On their own. Meanwhile even plants that cost the sweat of our
        brow,
Although in leaf across the land and all in flower now,
Either the sun will scorch them, sending too much heat below,
Or sudden cloudbursts and the icy hoarfrost lay them low,
Or they are hammered by a storm when violent whirlwinds
        blow.

Also, what's the reason Nature multiplies and feeds
The enemies of man on land and sea – the bristly breeds
220 Of wild brutes? How does it come about Disease abounds
At the change of seasons? Why does Death make his untimely
        rounds?
A human baby's like a sailor washed up on a beach
By the battering of the surf, naked, lacking the power of speech,
Possessing no means of survival, when first Nature pours
Him forth with birth-pangs from his mother's womb upon
        Light's shores.
He fills the room up with his sorrowful squalls, and rightly so! –
Just think what lies in store for him, Life's full supply of woe.[8]
But herds and flocks and droves of animals, both tame and wild,
When growing up, do not need rattles like a human child;
They have no use for lullabies, they do not need to be
230 Coddled with the baby-babble of the nursery.
They do not need a change of clothes according to the season,
And last of all, they have no need for weapons and no reason
For towering walls to keep belongings safe. The very Earth
And Crafty Nature bring forth all they need; they know no
        dearth.[9]

In the first place, since the stuff of earth, and water, and the fleet
Breezes of the wind, and also burning warmth of heat[10] –
Those elements we see comprise the Sum of Things – since they
Are made of substance that is born and that must pass away,

We must conclude the nature of the whole world is the same.
For clearly when we see the parts and limbs that make the frame   240
Of something are constructed of a born body that dies,
We know the whole is also born and perishes likewise.
Thus when I see even major members[11] of the world consumed
And born anew, then earth and heavens, it must be assumed,
Also have a birthday, and in time to come are doomed.

But in case you think that I am begging the question when I say
That earth and fire are mortal, and are bound to pass away,
Or when I have asserted confidently it was so
That air and water perish, and arise again and grow –   250
Let me explain. A good share of the earth, baked with the heat
Of the unrelenting sun, or trodden under throngs of feet,
Breathes out a cloud of dust and shifting veils of sand that fly
Scattered abroad by high winds to all corners of the sky.
And a share of soil is summoned back, washed off into the flood
With rains, and gnawing rivers eat away their banks of mud.
Besides which, everything Earth nourishes and makes to grow
Is returned accordingly. Because the Earth, as all men know,
Is not just the All-Mother, but the common graveyard too.
And thus you see Earth dwindles; but increased, she grows
    anew.   260

Furthermore, how ocean, springs and rivers ever abound
With new water, endlessly flowing, I need not expound –
The rushing down of waters all around us makes it plain.
Yet even so, the water levels, overall, remain
The same, since older water is taken away – a quantity
Skimmed by strong gusts of wind that scour the surface of the
    sea,
Some ravelled by the shuttles of the sun in heaven's vault,
And some spread underground through all the earth, so that the
    salt
Is filtered out[12] and the cleansed waters flow afresh and meet
At the fountainhead of rivers, whence the fluid new and sweet   270
Wells up and marches in a file across the land once more
Along the path its liquid feet went trooping down before.

Now I shall talk about the air, whose body never stays
The same a single hour, but alters in innumerable ways.
For things are ever flowing away, their streams of matter pour
Into the air's vast ocean; but if air did not restore
Matter to these objects, keeping them in good repair,
All things would, by now, have disappeared into thin air.
Therefore air never stops being produced from things, and going
280 Back into things, since everything's continuously flowing.

Again, the sun, that welling spring of liquid light on high,
Continuously floods fresh radiance throughout the sky,
Supplying new light in the place of old without delay.
For light's vanguard is ever lost and vanishing away
Wherever it falls. The following phenomenon will make
This clear: when clouds first slip between us and the sun, and
       break
The beams of light, the interrupted rays are lost below
At once, and the clouds plunge the earth in shade wherever they
       go.
290 Thus you can know that things require an ever-renewed supply
Of light, because wave after wave the previous shafts die,
And the only way things in the sun appear before our sight
Is they are ever replenished by that fountainhead of light.

Again, consider here on earth those lights we burn at night,
Hanging lamps, and brilliant, leaping torches, their flames rich
And thronging with thick billowings of smoke as black as pitch,
A supply of fire fuels these lights in exactly the same fashion.
Rushing to provide fresh brightness, they keep on and on
And on, with flames a-flicker, so that the space never falls dark,
300 And the chain of light is never broken, so quickly does new
       spark
Arise, to hide the quenching of old light, from fire's store.
We must conclude the sun, the moon, the stars above, therefore,
Also keep welling up with shining, and shedding it anew,
As flash after flash is lost forever. Thus you should not construe
That heavenly bodies are immune from ruin and decay.

Haven't you seen that even rocks are conquered by Time's sway,
That lofty towers are brought low, and stone crumbles away,
That the gods' shrines and statues weaken with wear and tear,
    and crack,
Nor can their holy power push the deadline of Fate back
Or struggle against the covenants of Nature? Don't we stare   310
To see the monuments of *men* in ruins? and writ there,
Such legends: 'Look upon my works ye mighty, and despair!'[13]
Don't we see boulders tumble down, wrenched from a mountain
    peak,
Unable to withstand what even *finite* time can wreak?
For the boulders would not tumble down, suddenly torn away,
If for eternity they had endured the whole array
Of Time's bombardments without crumbling.

                        Now take the case
Of sky, above, around, holding all earth in its embrace.
If it produces all things from itself, as some maintain,
And when things are destroyed, takes them back to itself again,   320
Then it is all of stuff that's born and dies. For whatever feeds
And nourishes other things out of its own body, must needs
Diminish, and when it takes things back, renew with their supply.

Besides, were there no birthday for the earth and for the sky,
If they'd existed always for eternity, then why
Is it in epic song the ancient poets never employ
Events before the *Seven Against Thebes* or *Fall of Troy*?[14]
Why have so many deeds of men just dropped into the gloom
And grafted on the eternal monuments of fame no bloom?
But this world's still a fresh and newborn thing, I hold it so: 330
The genesis of the world was not so very long ago.
And that's why, in the arts, we're making great strides to this
    day –
Ships were recently much improved. Nor was it far away
In the mists of Time musicians learned to make mellifluous song.
These very revelations about the universe are young,
And I myself the first to put them in our mother tongue.

But if you think these things were all invented before, and lost
When earlier races of men were burnt up in a holocaust,
340 Or a mighty convulsion of the earth laid all their cities low,
Or relentless downpours made the hungry rivers overflow
And inundate the land, engulfing towns beneath their spate,
All the more reason you must admit defeat – you demonstrate
Yourself that earth and sky are doomed to die! – since when
     such great
Dangers and infirmities have beset a thing, then all
It would take would be one even graver shock to make it fall
To pieces, dealing disaster far and wide. How would we know
That we ourselves are mortal, save that we see each other grow
Sick from the same diseases to which we've known others fall
     prey,
350 Whom Nature from this life has already spirited away?

Besides, there are three types of things that last forever: those
That being utterly solid in their substance shrug off blows
And which prohibit anything to penetrate inside
The close-knit fabric of their parts to rend them and divide –
Atoms are of this type, as I have shown you not far back.
Or else the reason things can last forever is they lack
Anything to *do* with blows, such as Void, which is not
Affected since blows cannot ripple emptiness one jot;
Or else it is because there is around them no supply
360 Of space into which their parts are able to dissolve and fly,
Just as the universe, the sum totality, will last
Forever since there is no other place beyond and past
The universe for things to leap to, or place whence there could
     come
Hard blows to rain upon it and to shiver apart the Sum.

But, as I've shown, the world is not all solid; it's alloyed
With emptiness, for things are mixed of matter and of void.
Nor is it like a vacuum; bodies can by chance be swirled
Together out of the Infinite to lay waste to this world
In a savage cyclone, or import some other deadly doom.
370 Nor further, do the vast reaches of Space lack any room

Into which the ramparts of this world could fly and scatter –
Whatever kind of force they are assaulted by to shatter.
And thus the Door of Death is neither shut against the sky,
Nor bolted against the sun nor earth nor watery deep, but yaws,
And hungrily awaits them with its huge and gaping jaws.

So you must own these things[15] were also born. Things which
    consist
Of substance that is mortal couldn't have shrugged off and
    dismissed
For all the ages up to now the onslaught of Time's might.
Next, since the major members of the world[16] so fiercely fight   380
Amongst themselves, stirred up in an unholy civil war,
Do not you see that their long clash can't last forever? For
Perhaps the sun with his host of heat will gain the upper hand,
Having dried up all the waters, a conspiracy long planned.
But he has yet to carry out his plot, for the supply
Of the rivers is so abundant, these not only don't go *dry*,
They threaten to drown all from the whirlpool of the deep – in
    vain –
Since winds in *their* turn scour and skim the surface of the main.

And the waters also shrink beneath the towering sun who frays
The surface, unpicking the fabric with the needles of his rays,
While the rays have faith that they can dry the floods before the
    seething   390
Waters accomplish their own ends – so fiercely these
    war-breathing
Elements contend in well-matched battle to decide
The conflict and the vital matters of the world that ride
Upon it. Though in the turmoil, once Fire gained the upper
    hand,
And once, so legend has it, Water lorded it over the land.

For Fire won the day, and circling the globe, it seared
Many lands, when the headstrong horses of the Sun careered,
Dragging Phaëthon all round the earth, throughout the sky.
But the All-Father, goaded to anger, suddenly let fly

400 A thunderbolt that dashed foolhardy Phaëthon from on high,
Out of the chariot, down to the ground. Meeting him as he
      hurled
Earthward, the Sun caught up the eternal lantern of the world,
And gathered together the bolted team, re-yoked each
      shuddering horse,
And restored the world again, steering the steeds back on their
      course.
Or so sang the Greek poets of yore, but in reality
This legend is as far removed from reason as could be!
For Fire gains the advantage when the bodies of that stuff
Rise up out of the Infinite in numbers great enough,
But then its ranks fall back some other way and meet defeat;
410 Otherwise, things perish, burnt up in the blasts of heat.

Water also started to prevail once, legends tell,[17]
When it submerged much of the human race beneath its swell;
But then when all the water's force, which had been drummed
      up out
Of the Infinite, was somehow turned aside and put to rout,
The rains came to a halt, the rivers checked their rush of water.

But next I shall set forth, in order, how this mass of matter
Laid down the groundwork for the earth, the sky, the briny
      deep,
And also for the courses that the sun and moon must keep.
For obviously the primary particles did not scheme to fit
420 Themselves each in their proper order by their cunning *wit*.
Nor did they strike a deal amongst themselves exactly how
Each should move. Rather, for time infinite up to now
Myriad primary particles moving in many directions, whether
Driven by blows, or their own weight, were wont to come
      together
Every which way and experiment with every permutation
And everything that they could fashion by their combination,
And as a result, the particles, spread out over a vast
Span of time, by trying each movement and combination, at last

Suddenly hit upon the combinations that can be
The building blocks of greater things, the earth, the sky, the sea,   430
And all the generations of living beings.

                                There was no sight
At that time of the sun's high-flying wheel, lavishing light,
Nor was the far-flung firmament spangled with galaxies,
Nor was land to be seen, nor air, nor oceans, nor the skies,
Nor anything at all like those things we would recognize,
But a strange and swirling maelstrom, a gathering mass of every
     guise
Of atom, Strife[18] stirring skirmishes up amongst them, and
     confusion,
A mêlée of different intervals, of ways, connections, motion,
Weights, combinations, blows – because, due to the manifold   440
Shapes and forms of atoms, not every group of them could
     hold
Their configuration or make motions that could work together.
Then like began to join with like, parts started to dissever
From the chaos and draw the boundaries of the world, to
     organize
Its members, setting its mighty parts in place: the towering skies
Divided from the land, the open ocean set in turn
Separate and apart from where the aether's pure fires burn.

And this is how: first, particles of earth began to gather
In the nucleus, since they were heavy and tangled up together,   450
And earth all hunkered down into the deepest spots. The more
Tightly knit the earth, the more it squeezed out from the core
The bodies which would build the sun, the moon, the stars, the
     seas
And the vast ramparts of the world, because the seeds of these
Were light and round, much smaller than the earth's. Therefore,
     the first
Element to escape through sparse gaps in the earth and burst
Forth from it was fire-bearing aether, climbing to a height,
And carrying away much of the fire[19] with it, being light.

460 Not unlike a scene we've often had the chance to view –
When, in the early morning, above the grass bejewelled with
        dew,
The day breaks, and the golden radiance of the sun is kissed
With red, and lakes and year-round-flowing streams breathe out
        a mist,
So that sometimes it seems the earth is steaming. Then on high,
The evaporations gather up beneath the vaulted sky
And knit a scrim of cloud-cover. This selfsame method served
The aether, which though thin and flimsy, wove one cloth that
        curved
And spread out far in all directions, blanketing every place,
470 Encircling all else within in its amorous embrace.

Then after came the web on which were woven moon and sun,
Whose courses in the air between the earth and aether run,
For they were such that neither of those elements would adopt
Them as their own, since they weren't heavy enough so that they
        dropped
Down and settled, nor were they light enough that they could
        glide
In the uppermost reaches – so they between the two extremes
        reside,
And spinning like living bodies, as members of one world, abide.
(The same is true with our own bodies, for we can attest
Parts of us can move while other members stay at rest.)
480 When these orbs were delved out from it, the earth suddenly
        gave
Way where now blue billows of the open ocean lave,
And flooded the basins with the swirling brine. And day by day
As blasts of surrounding aether and the sun's rays pounded away
At the outer crust of the earth from every side, they packed it
        close
And compressed it. And the more it was pressed together by the
        blows,
Collapsing in upon its own centre, the more it squeezed
Out of its flesh the salty sweat whose seeping brimmed the seas

And the swimming savannas of water, and the more the
    quantities
Of particles of heat and air escaped its mass to fly   490
Far over the earth and crowd the gleaming precincts of the sky.
The plains then settled, putting in relief the towering crown
Of mountains – for rocky ridges were not able to sink down,
Nor could the crust in every place subside and settle the same.

And in this fashion the massive body of the earth became
Solid; and all the 'silt' of the cosmos filtered down, as dregs
Sift down beneath their weight and gather at the base of kegs.
And then the sea, the air, the aether with its fires that burn –
All of them fluid substances – were left pure in their turn,
As each was lighter than the last. The lightest of all these   500
And most fluid being aether, which soars above the airy
    breeze,
And does not mingle its pure flowing with the storm below
Of clashing winds, but leaves the rest to be tossed to and fro
By raging cyclones, at the mercy of any gales that blow,
While, carrying its fire, it glides on steady and serene.
That something can glide steadily one way like this we've seen
In the case of the Black Sea, whose currents uniformly glide
In one direction with no fluctuations in its tide.[20]

And now the reason for the movements of the stars I sing.
First of all, if it is from the heavens circling
As a great sphere, then we must venture that the heavens roll   510
Due to an outside air that pinches it at either pole,[21]
While yet another flow of air pours from above, to spin
The sphere in the direction that the stars are moving in,
Twinkling in the eternal firmament. Or else there's bound
To be another current heaving and lifting the world around
From below, as rivers turn the paddles of a water wheel.

It's also possible the entire firmament stays still
And it's the shining star-signs that are moving – whether from
Surging waves of aether trapped within the heavenly dome,

520 That, whirling round, seeking an exit through which they can
        fly,
    Roll the fiery balls throughout the thundering night sky,
    Or a wind from *beyond* the world blows them around. Or else it
        might
    Be that flocks of stars go wandering where appetite
    Entices each to go, so they meander as they graze
    The pasture of the heavens, feeding bodies all ablaze.

    But which of these is the true cause, it's hard to ascertain.
    Rather, it is the *possibilities* that I explain –
    What things can and do come about in all the universe
    In the many worlds created different ways. I give divers
530 Rationales which can explain the motion of the stars
    In *all* the worlds – and *one* of these has to hold true for ours,
    Empowering stars with motion. Which is right? We cannot say,
    When we are only blindly, step by step, feeling our way.

    The reason that the earth can lie at rest right in the middle
    Of the world is that its density must lessen little by little
    And its weight lighten by degrees, and underneath, the earth
    Must be of another substance, though united at its birth
    With the airy parts on which it lives engrafted; hence the weight
    Of the earth does not load down the air, any more than limbs
        are freight
540 To the person who possesses them. Neither is the head
    A burden to the neck, nor is the body so much dead
    Weight that we feel it bearing down upon our feet. Instead,
    It is the weights that come from *out*side of ourselves that gall
    When placed upon us, though often relatively very small.
    What matters in the end is what each has the power to do.
    And thus the earth is not some foreign object placed onto
    An alien air from elsewhere. Rather, from the very start
    Of the world they were conceived together, each an intrinsic
        part,
    Just as our own limbs are to us. Thus, when a terrific bolt
550 Of thunder strikes the earth all of a sudden, then the jolt

Also shakes the air and everything lying above the ground.
And this event could not occur unless the earth were bound
To the skies and airy regions of the world – all combined
And tangled up together by roots shared and intertwined,
United from the dawn of their existence. Don't you see
That though the soul is of a stuff as flimsy as can be,
Yet it can shoulder the hefty bulk of our frame – because it
    weaves
With the body into one whole cloth? And what is the force that
    heaves
The flesh up in a nimble spring, if not the spirit's might
Commanding the legs? Now don't you understand how such a
    slight   560
Substance can be powerful when once it is combined
With a heavy mass, as the air joined to the earth, and as you'll
    find
Is true for us when flesh is teamed up with the power of mind?

Next, the heat of the sun and the diameter of its wheel
Must be roughly the same size it appears and our senses feel.
For over whatever distance fires are able to shed their glow
And blow their heated breath upon our limbs, yet even so,
Over this span, they do not lose a jot of their mass of flame;
Rather, the size of the fire to our sight remains the same.
Therefore, though the heat of the sun and its outpouring of
    light   570
Reach our senses, making all the regions of earth bright,
The magnitude of the sun appears so truly to our eyes,
That this must be, with no exaggeration, *actual size*.

As for the case of the moon: whether it is with her own beams
Shed from her body somehow that she bathes the earth, or
    gleams
With bastard brightness, either way, the magnitude we see,
As she travels before our eyes, is just the size that she must be.
For all things we remark at great remove, through a thick scrim
Of air, before they shrink, their outlines first go blurred and dim,   580

But since the outline tendered by the moon is sharp and clear,
Her size and shape must be precisely as they both appear –
The circumference in the sky the same as we perceive it here.

Lastly, with all the heavenly fires you see, it is the same.
For instance, take the earthly fires we view. As long as flame
And flickering are clear, as long as we see them burn, how far
Away they lie, as a rule, scarcely affects the size they are
To our sight, one way or the other. And the same's true for a
     star:
590 *What you see is what you get*. If their size differs a mite
From their appearance, the degree is trivially slight.

Another matter that should not bewilder you at all
Is how the sun can give off so much light if it's so small,
Such it can make the sea, and all the lands, and skies replete,
Soaking them with light and inundating them with heat.
It's possible that from this spot wells up the artesian spring
Of light for the whole world, gushing forth and scattering
Its full flood of sunshine, due to heat-elements that course
600 Together from all quarters of the world, and from which source
The heat pours forth as from one fountainhead. Haven't you
     seen
How, at times, a meagre spring of water can drench the green
Breadth of a meadow and flood the fields?

                                        For even if the blaze
Of the sun is not so great in magnitude, its burning rays
May yet lay hold of the air, if the nature of the air is such
That it's combustible and ready with the slightest touch
Of fire to catch alight, even as we often mark
That in a field of grain or stubble it takes a single spark
For everything to go up in flames and the whole field to scorch.
610 Or else perhaps the sun can glow on high with rosy torch
Because the sun possesses round himself a recondite
Ring of fire, undetectable to human sight,
But whose heat greatly magnifies the force of visible light.

Nor is there one neat explanation how the sun is borne
From summer to the turning point of wintry Capricorn,
And rounding the bend, how he comes back, making for the
  post
Of Cancer's summer solstice – or how the moon is seen to coast,
In the space of only a month, the very distance that the sun,
Driving along his track, takes an entire year to run.
Again, there are many possible explanations, not just one.   620

Among the most plausible reasons is the one that was put forth
By wise Democritus in his scriptures – the closer to the earth
A celestial body is, the less it can be blown along
By the whirlwind of the heavens, for the current blows less
  strong
The lower down it is, and its swift power flags below.
And thus the sun, compared to the star-signs, is travelling slow,
Being lower than the fiery constellations, which one by one
Come from behind and then gain on and overtake the sun.
And more so with the moon, for since the moon hangs even
  lower,
Closer to the earth and further from the sky, she's slower
And is less able to catch up with the star-signs on her course.   630
And as the moon is blown along by wind of weaker force
And is lower than the sun, the cycling star-signs overtake
The moon that much more rapidly, and leave her in their wake.
And that is why the moon seems to be moving backwards faster
To each sign; really, star-signs are more quickly gliding past her.

Another possibility is two air-currents flow
Crosswise through the heavens, and they alternately blow
At certain times: one that's able to push the sun away
From the summer constellations to midwinter's shortest day   640
And the locking up of ice; and another to return
Him back from the chill gloom to where the summer star-signs
  burn.
And we must also assume that the same mechanism steers
The moon and planets in wide orbits over lengthy years:

They travel due to other airs that alternately flow.
Haven't you seen how clouds caught in opposing winds will go
In opposite directions – the top against the layer below?
So why can't celestial bodies through vast rings of aether glide,
Driven by the flow of this or that contrary tide?

650 When night drowns earth in its prodigious gloom, the reason
        why
Is either that the sun has reached the limit of the sky,
And worn out from the race, gasps out his fire with none to
        spare,
Shaken and made frail from travelling through so much air,
Or else because he's made to bend his course beneath the ground
By that same power that over earth propelled his sphere around.

Likewise Matuta at a certain hour spreads the rose
Blush of dawn across the shores of heaven so it glows,
Either because the very same old sun, on his returning
From underground, shoots rays ahead to set the heavens
        burning,
660 Or else because the fires gather and many heat-seeds stream
Together at a certain time, and each day make the gleam
Of a new sun afresh. And this phenomenon, some say,
Can be beheld on Ida's towering peaks at break of day –
For fires first scattered across the ridges then start to cohere
Into a sort of ball, until they form a single sphere.

But it should not surprise you that the seeds of heat can flow
Together at a certain hour to restore the glow
Of the sun; because we see that this is so for many things –
They happen at fixed intervals. It is one season brings
670 The orchards into bloom, while at another time the trees
Shower their petals down. No less for us: Nature decrees
The age at which our baby teeth fall out, or youths begin
To sprout a fuzzy down upon the smooth skin of the chin,
And when upon the cheeks a velvety beard begins to grow.
Then last of all, take weather – lightning bolts, rain storms,
        wind, snow –

These events do not occur at any random season.
The initial pattern of the web of causes is the reason
That from the world's beginning, things fell out in just this way,
And why they come back round at certain intervals to this day.

As to why the length of days grows longer while the night				680
Wastes away, or why the night increases, while the light
Dwindles, the explanation is either that the selfsame sun
Whose courses both above the earth and underneath it run,
Slices the sky's regions into unequal arcs, so that
The length of his orbit over and underground are disparate,
And whatever he subtracts from one leg of the trip, he will
Add that amount onto the other leg of it, until
He arrives in either star-sign of the year where equal room
Is allotted to the light of day and to nocturnal gloom;
For, midway on the courses blown by the north and the south
	wind,
The goalposts of the heavens stand equidistant at either end			690
Due to the position of the whole round zodiac
Through which the sun goes crawling to complete his yearly
	track,
Drenching heaven and earth with slanting light – so this event
Is explained by those who chart the quarters of the firmament,
All star-signs fitted into place. Or else perhaps it might
Be that the air is denser in some parts, and the flickering light
Of the sun's fire is bogged down underneath the earth, and slow,
And cannot break through easily, burst forth, and rising, glow;
And that's why winter nights drag on and loiter in this way
Until dawn's shining standard heralds the coming of the day.[22]		700
Or it is possible, if what some people say is true,
And fires gather and cohere to make the sun anew
Each day in the east, that it amasses for some reason
Quicker or more slowly in accordance with the season.

And as for lunar phases, it might be the moon sheds light
Struck by the sun's rays, turning that shining towards our sight
More and more each day as she withdraws from the sun, until
She faces him directly and her shining's bright and full.

And as she rises, lifted up to lofty heights, she sees
710 His setting, from which point she must conceal, by slight
    degrees,
Her light behind herself, the closer to the sun she flies
Through the bowl of constellations from the far end of the skies:
That's what they claim who think the moon's shaped like a ball,
    and runs
Her course along a pathway that lies underneath the sun's.

Or it might be instead the moon can shine in her own right,
And as she turns can give off different phases of her light
Because there is another body gliding alongside,
Obstructing her light in various ways, but which can't be
    descried
By us because it does not shine with any light at all.

720 Another possibility: she turns round like a ball,
Only one half of which has been dipped in a radiant glow,
And it is by revolving that the different phases show,
Until the hemisphere that is endowed with a fiery blaze
Is turned towards our open eyes and fully meets our gaze.
And then she turns this face behind, removing from our sight
Little by little the surface of the ball that gives off light,
Or so the teaching of the Babylonians[23] avers,
Seeking to disprove the science of astronomers.
(As if the champions of both theories did not have their case,
730 Or there were any reason one should hazard to embrace
One theory over the other!)

                        Indeed why couldn't it be true
That each of the lunar phases is a fresh moon fashioned new,
And that each new-created moon then disappears each day,
And another's formed to take the place of one which passed
    away?
You'd have your work cut out for you to prove it can't be so,
When so many other things arise in order, as you know:
Spring and Venus come, and wingèd Cupid leads the way,
And Mother Flora, in Zephyr's footsteps, scatters a bouquet

Of blossoms coloured bright to catch the eye, and smelling
    sweet,
Carpeting the path that stretches out before their feet.   740
Scorching Heat comes next, with Ceres head to toe in dust,
Attended by the Northerlies that in the summer gust;
Next in the annual procession, we see Autumn traipse,
Swaying arm and arm with Dionysus, lord of grapes,
Then other seasons follow, and the other winds blow past:
Volturnus the sky-thunderer, the South-wind's lightning blast.
And then, with snow and frost in tow, the Solstice of the year,
While Winter, his teeth chattering with cold, brings up the
    rear.[24]
Thus it is not so strange that a fresh moon should come about
On certain days, or, at specific times, be blotted out,
Seeing that so many things arise at a fixed season.   750

Again, you must consider there is more than just one reason
That explains the skulking moon and the eclipses of the sun.
For why should the moon be able to rear her head to such a
    height
She stands between the earth and sun and so blocks out his light,
Casting her darkened sphere right in the way of his shining
    flame?
Why not instead believe another body is to blame,
Forever at her side gliding along, forever dark?
Why shouldn't it be possible the sun could lose his spark
And gutter out at certain times (and at fixed times renew),
When, passing through the air, he crosses regions hostile to   760
His blazes, areas that cause his fires to cease to burn?

And then, why should the earth have the ability, in turn,
To steal the shining from the moon, by getting in the way,
Passing above the sun and keeping him beneath its sway
As the moon along her monthly course cuts through the shady
    cone?[25]
Why could not another body be the explanation –
Running underneath the moon, above the sun's sphere, so
Interrupting his beams of light and cutting off their flow?

But if the moon can shine under her own power, then too
Shouldn't it be possible that when she passes through
770 Parts of the heavens hostile to her light, it is snuffed out?

But next, since I've explained how everything can come about
Under the sweeping blue of heaven, so we can know what forces
And reasons propel the sun and moon along their different
        courses,
And how it is their light can be obstructed, how they're made
Invisible, how they shroud benighted lands beneath their shade,
Then how they wink and open up again and turn their gaze
Upon all places, bathing them in brightness of their rays,
780 Now I return to the dawn of the world, and soft fields of the
        earth,
And what they brought forth on the shores of Light in that new
        birth,
Abandoning them to the shifty breezes.

                                          First, earth cloaked the scene,
Hill and dale, with every kind of leaf and shining green,
And green the blooming meadows gleamed. All trees began to
        vie
Galloping at a terrific clip, to race up towards the sky.
And just as feathers, fur or bristles straightaway start to grow
On four-footed beasts or on birds mighty-on-the-wing, just so
790 The fledgling earth first sprouted a down of herbs and coppices.
Next, she engendered the tribes of living things, the many races
That arose by different causes and in many different ways.
For beasts could not have dropped from the sky, nor on the
        other hand
Could they have crawled out of the briny pools to walk the land.

Only Earth is left, therefore, deserving of the name
Of Mother, since it's from the Earth all living beings came.
For there are many creatures even now that spring to birth
From soil that's sodden by the rain, coddled by sunny warmth,

So it's no wonder more and larger creatures at that time
Sprang from the soil when Earth and Air were still fresh in their
    prime.   800
First of all, the pinioned nations, in the world's first spring,
Birds of every species, hatched from eggshells and took wing,
Just as now cicadas in the summertime divest
Themselves of their shiny coats, and go out in the world in
    quest
Of life and livelihood. It was at this time that the Earth
First brought forth the kinds of beasts, for then there was no
    dearth
Of warmth and moisture in the fields. Where purchase could be
    found,
Wombs began to take root, and to spring forth from the
    ground,
And when the embryos had reached full term, and burst from
    there,
Escaping from the watery sac and gasping for the air,   810
Then Nature channelled the Earth's pores, and made the opened
    veins
Flow with a sap akin to milk, as after labour pains
A mother wells with sweet milk, because nourishment in a flood
Is pulled towards her breasts. Earth gave sustenance to her
    brood,
Warmth clothed them, and the herbs and grasses offered them a
    bed
Thick and soft and deep with down to rest a weary head.
But the dawn-time of the world did not bring forth harsh cold
    and snow,
Nor too much heat nor mighty winds that violently blow,
For all things in the same way become stronger as they grow.   820
Therefore, again and again I tell you, that when men acclaim
The earth as 'Mother Earth', she is deserving of the name,
Since it was she made man, and at fixed times, made every
    other
Tribe of beast that roisters across the mountainsides, together
With the winged creatures of the air in every form and feather.

But since there must be an end to bearing, so too with the
    Earth –
Like a woman wasted with long years, she left off giving birth.
For Time changes the nature of the whole world, and one phase
830 Must be succeeded by the next; there is no thing that stays
The same. Everything flows. Nature makes everything alter,
For as one thing grows feeble with old age and starts to falter,
Another strengthens, emerging from obscurity. So Age,
Therefore, changes the nature of the whole world, and one
    stage
Of the Earth gives way to another; she cannot bear any more
What once she could, but now brings forth what she could not
    before.

In the beginning, there were many freaks. Earth undertook
Experiments – bizarrely put together, weird of look –
Hermaphrodites, partaking of both sexes, but neither; some
840 Bereft of feet, or orphaned of their hands, and others dumb,
Being devoid of mouth; and others yet, with no eyes, blind.
Some had their limbs stuck to the body, tightly in a bind,
And couldn't do anything, or move, and so could not evade
Harm, or forage for bare necessities. And the Earth made
Other kinds of monsters too, but in vain, since with each,
Nature frowned upon their growth; they were not able to reach
The flowering of adulthood, nor find food on which to feed,
Nor be joined in the act of Venus. For all creatures need
Many different things, we realize, to multiply
850 And to forge out the links of generations: a supply
Of food, first, and a means for the engendering seed to flow
Throughout the body and out of the lax limbs; and also so
The female and the male can mate, a means they can employ
In order to impart and to receive their mutual joy.

Then, many kinds of creatures must have vanished with no
    trace
Because they could not reproduce or hammer out their race.
For any beast you look upon that drinks life-giving air,
Has either wits, or bravery, or fleetness of foot to spare,

Ensuring its survival from its genesis to now.
And many other animals survive because of how   860
Useful they are to us – entrusted to the care of Man.

First, take the savage race of lions, that ferocious clan –
They've been preserved by courage; foxes by craftiness; stags,
    flight.
But the dog, steadfast of heart, whose sleep is vigilant and
    light,
And every kind of animal of beast-of-burden bred –
The sheep of woolly fleece, and oxen of the hornèd head –
We've taken under our protection, Memmius, all of these,
For they were happy to flee from predators, and longed for
    peace,
And to feast on food not got by their own labour at our board –
Which for their usefulness we offered them as their reward.   870
But those Nature did not endow with suchlike qualities
Could neither make a living by their own efforts, nor please
Us with some kind of useful service by which they could reap
The benefit of our protection and could earn their keep.
So it was open season on those brutes, for prey or gain
By others, hobbled utterly by their own doomful chain
Until Nature finally drove their species to extinction.

But there were never Centaurs, nor were there at any point
Monsters of twin natures, with a twofold body joined
Together out of mismatched limbs, such that the natures both   880
Could be compatible in powers, equal in their growth.
And this should convince anyone, however dim or dull:
First, the horse, at three years old, has reached the pinnacle
Of his strength. But a three-year-old boy? Not at all. For in the
    night
He still dreams of suckling at the breast. Then when the might
Of the horse deserts him in old age and his legs lose their power
With the ebbing of life, that's when the youth is just reaching the
    flower
Of manhood; on his cheeks, a velvety down begins to grow.
In case you still had any doubts, this evidence must show

890 That Centaurs can't exist, composed of Man and of the seed
     Of the heavy-laden horse, nor likewise can there be a breed
     Of Scyllas, all snarling dogs about the middle, yet combined
     With the tail of a giant fish, nor other monsters of this kind,
     Made of a mix of hodgepodge limbs, which do not reach the
          prime
     Of maturity, nor reach the fullness of strength at the same time,
     Nor cast it off again at the same point in old age, nor burn
     With the same lust, nor have behaviour compatible in turn,
     Nor are the same things good for them for which their bodies
          yearn.
     Take hemlock – bearded goats graze on it and grow sleek and
          wide,[26]
900 When it's a toxic plant to us; in fact it's suicide.

     Again, since the tawny body of the lion is just as much
     Susceptible to singeing and to burning at the touch
     Of fire as any other thing of flesh and blood we see,
     Then how could a Chimaera with one body made of three –
     A lion at the fore, a dragon at the rear, a goat
     Between – breathe scorching flame out of its flesh and through
          the throat?

     Therefore he who imagines that the world was able to
     Give rise to creatures of this sort when earth and sky were new –
     Resting his whole argument on 'new', that empty word –
910 In the same vein, may babble hogwash equally absurd:
     He may claim that, in those days, golden rivers used to flow
     All over the face of the earth, or that the boughs of trees would
          glow
     With gems instead of blooms, or there were men who, with one
          leap,
     Because of the giant span of their long limbs, could cross the
          deep
     And juggle overhead the whole of heaven with their hands;
     For even though many seeds of things abounded in her lands
     When first the Earth produced the animals, that is no sign
     That she could cobble together hybrid creatures that combine

A jumble of odd parts into one whole. Varieties
Of plants, moreover, fields of grain, and the rejoicing trees,   920
Things that to this day come teeming from the earth, you'll find
Still cannot be created of two species intertwined,
But everything matures after the manner of its kind,
And every breed by Nature's settled covenant maintains
Its individual traits.

                    But back at *that* time, on the plains,
The race of humankind was far more hardy, as befit
The very hardness of the earth that had engendered it.
And they were built on scaffolding of bigger, denser bone,
Fixed with brawny sinews throughout the flesh, and weren't as
    prone
To being overwhelmed by heat or cold, could stomach all
Kinds of changes in their diet, and they did not fall   930
Ill from any sickness. For many a cycle of the blaze
Of the sun rolling through the heavens, they dragged out their
    days
Like the far-roaming wild brutes, nomadic in their ways.
Back then there was no sturdy ploughman to guide the curving
    plough,
No one knew how to work the land with iron tools, or how
To plant young slips in soil, or cut the barren branches down
From the tall trees with pruning hooks. Whatever sun and rain
Provided them, whatever the earth, unasked for, would impart,
They found these things were boon enough to satisfy the heart.

Mostly they would take a mess of acorns for their meat
Amongst the groves of oak, or from arbutus they would eat   940
The berries – which you see are just now in the wintertime
Ripening to scarlet. But when the Earth was in her prime,
She bore more generously, and larger things besides. Back then,
When the Earth was in her first bloom, she provided wretched
    men
With many foods, more than enough to eat, though rustic fare.
And springs and rivers beckoned them to slake their dry throats
    there,

As now tall waterfalls cascading down the mountainside
Thunderously call the thirsty creatures far and wide.
Then men took up the wooded haunts of nymphs, places they'd
    found
During their wanderings, where they knew water to abound,
950 Trickling from a spring and slipping over stones, across
The slippery damp stones, and dripping on the verdant moss,
Welling up here and there, and on the broad plain widening out.
They did not know how to treat things with fire, or know about
The use of hides, or how to dress in skins despoiled from kills.
They dwelt in glades and forests and in caverns in the hills.
When lashing wind and rain made them seek shelter from the
    sky,
They hid their dirt-caked bodies under thickets to keep dry.
They could not look out for the common weal. There were not
    then
Laws or customs governing the ways men dealt with men;
960 But each man seized what plunder Chance put in his way. To
    thrive,
Each learned to watch out for himself, his own will to survive.
Then Venus wedded the bodies of lovers in a sylvan bower.
Man won his mate by shared desire, or he would overpower
Her with his violent strength and lust – or wooed her with a
    treat
Of acorns and arbutus fruits, or fine ripe pears to eat.

And relying on the wondrous abilities of hands and feet,
They chased their quarry of forest animals, and felled their prey
With stones or cudgels. Many they slew – from some, they ran
    away
970 By fleeing into their lairs. And like the wild and bristly boar,
When night caught up with them, they lay down on the forest
    floor
Rolling their naked, woodland-dwelling limbs up in a nest
Of foliage and the boughs of trees.[27] They did *not* go in quest
Of the vanished sun when day was done, shivering with fright,
Roaming the fields with loud lament through shadows of the
    night;

But quiet, tombed in sleep, they waited for the sun to rise
Again and with his rosy torch illuminate the skies.
For since they had been used to seeing the alternating change
Of light and dark since childhood, it could never have seemed
    strange,
Nor could it make them fear that on the earth, eternal night   980
Held sway, and that the sun would never come back with his
    light.
What worried wretched man instead was, when asleep, he lay
At the mercy of the tribes of wild animals, easy prey.
At times he fled, evicted from his rocky dwelling, when
A wild boar frothing at the mouth or powerful lion burst in,
And in the dead of night, roused terror-stricken from his rest,
He yielded up his leafy pallet to the savage guest.

Back then the fate of an untimely death was no more rife
Than now, when men with moaning leave the sweet light of this
    life.
To be sure, each was more likely to be caught by some wild
    beast,   990
Gulped down in toothy jaws, supplying it a living feast,
Filling the groves, the hills and woods with groans, because he
    was
Buried alive, he saw, inside a live sarcophagus.
And those who managed to escape, but with their bodies
    mauled,
Later placed shaking hands on suppurating sores and called
On Orcus with hair-raising cries, until the pains that racked
Their flesh released them from their lives, and all because they
    lacked
Aid and the know-how to dress a wound. But no one day would
    yield
At that time myriads of men reaped on the battlefield;   1000
Neither in those days did tossing surges of the main
Shiver ships and sailors on the rocks. Blindly, in vain,
To no purpose, often the sea would rage with rising tide,
Then fickle as you please, would toss her empty threats aside.

Nor then could the bewitching laughter of the sparkling waves
And peaceful-seeming sea beguile men to watery graves;
The perverse science of navigation still lay hid in gloom.[28]
Back then a dearth of food sent swooning bodies to the tomb;
Now men are sunk beneath excess and eat more than their fill.
Then, men unwittingly ingested poison that would kill;
1010 But now men poison others, being expert in that skill.

Then after acquiring shelter, hides and fire, man and wife
Were joined [and lived beneath one roof, and learned to share a
        life,]
And realized it was their union that produced a child.
That was when the race of man first started to grow mild.
Then fire saw to it that their shivering bodies could no more
Endure the cold beneath the vault of heaven as before,
And Venus drained their powers, and the little ones, with ease,
Broke down the stubborn pride of parents with their coaxing
        pleas.
Then neighbours began to form the bonds of friendship, with a
        will
1020 Neither to be harmed themselves, nor do another ill,
The safety of babes and womenfolk in one another's trust,
And indicated by gesturing and grunting it was just
For everyone to have mercy on the weak. Without a doubt
Occasional infractions of the peace would come about,
But the vast majority of people faithfully adhered
To the pact, or else man would already have wholly
        disappeared;
Instead, the human race has propagated to this day.

But it was Nature gave the tongue its different sounds to say,
And expedience that formed the names of things – much the
        same way
1030 We see infants are driven to point their finger and to reach
At what they want to show, precisely from their lack of speech.
A sense of what its powers are suited for is given to each –
The young bull, even before the horns have sprouted from his head,
Already tries to charge and butt with these when seeing red,

While whelps of panthers and the cubs of lions already fight
Tooth and nail when they have scarcely any fangs to bite
Or claws to scratch with yet. And we've seen how a fledgling
     flings
Itself into the air, trusting the wobbly aid of wings.   1040

And therefore to assume there was one person gave a name
To every thing, and that all learned their first words from the
     same,
Is stuff and nonsense. Why should one human being from
     among
The rest be able to designate and name things with his tongue
And others not possess the power to do likewise? Moreover,
If he had never witnessed others using words before,
Then how did the idea of speech first germinate and grow,
And where did he get a concept of its usefulness, to know
In his mind what he wished to do? He wouldn't be able to
     intrude
And force his will, a single man's, upon the multitude   1050
So that they wanted to adopt *his* names of things. It's clear
That it is difficult by any means to make men hear
What is needful to be done, when they turn a deaf ear.
For they would not have borne unheard-of utterances a minute,
Grating on their ears, if they could see no purpose in it.

Why should it be so wonderful the human race expresses
Different things and feelings with different sounds, since it
     possesses
Such marvellous instruments of tongue and voice, when if you
     take
Dumb flocks, and even all the creatures in the wild, they make
A range of sounds, and voices in a wide variety   1060
When they are terrified, or are in pain, or burst with glee?
This is quite obvious from observation, as you see.

The sound a provoked Molossian mastiff[29] utters when she
     growls,
Baring her sharp teeth by drawing back her hanging jowls

In fury, with a threatening snarl, is not at all the noise
She makes when filling the air with barks, baying at full voice.
Then when she tries to clean her puppies gently with her tongue,
And bats them with her paws, making as though to eat her
     young,
Grabbing and holding them gingerly between her teeth in play,
1070 The way she yelps when nuzzling them is not at all the way
She howls when she's abandoned, shut up in the house for
     hours,
Or shrinking from a kick, she whines and whimpers as she
     cowers.

Consider the different whinnies of a stallion: when youth stirs
His blood to frenzy, and Wingèd Love is goading him with
     spurs,
He makes an eager whickering among the herd of mares;
But when he's spoiling for the fight, he trumpets and he flares
His nostrils; other times he neighs when terror makes him shake.
Lastly, consider the feathered nations of the air – for take
Sea-hawks, ospreys, sea-gulls, all the kinds of birds that make
1080 A living among the billows of the sea, in the salt spray.
The cries they make while in the chase are not in any way
The noises that they make when squabbling over captured prey.
And then, with changes in the weather, some birds change their
     squawks –
As, for instance, the case of parliaments of rooks or flocks
Of long-lived crows, which folks claim make use of a certain call
To summon winds, and with another, ask the rains to fall.
Therefore if different feelings give even dumb animals no choice
But to make different utterances, consider the human voice,
How much more likely it was for Man with his wide range of
     sound
1090 To indicate and distinguish all the different things he found!

In case you're wondering to yourself about how fire first came
To mankind, lightning brought it down to earth. This is the
     flame

That is the source of all the other fire. For we descry
Many things ignite when seeded with fires from the sky
Once the celestial bolt imparts its heat. Yet when winds blow
And buffet a tree with blasts that set it rocking to and fro,
Its limbs rub up against the branches of its neighbour, so
That fire is forced out by the violent friction of the bark –
At times the burning heat of flames gives off a flash of spark
From the rubbing together of the trunks and branches of the
    trees.   1100
It's possible that men got fire from either one of these.[30]
And after that, the sun instructed them to cook their meat,
And how to make their victuals soft by using fire's heat,
Since they were used to seeing many a thing out in the field
Beneath the onslaught of the sun's hot rays grow mild and
    yield.

Then those who towered above the rest in intellect and bold
Imagination, day by day showed how to change the old
Manner of life with fire and with new ways of doing things.
And next, cities and citadels sprang up, founded by kings,
Who constructed these defences for their own protection and
Divided up among their subjects herds and plots of land,   1110
Allotted according to beauty, strength or intellect. (In those
    days,
Beauty was held in high esteem, and muscle won great praise.)
Then later on, wealth was invented; men discovered gold,
Which easily stole from fair and strong all rank that they might
    hold.
(For a man will follow the retinue of a richer, as a rule,
However strong in frame he is himself, or beautiful.)

But if you'd steer your life by a philosophy that's true,
The way to be the wealthiest of men is to eschew
High living, and be contented in the mind – for there has never
Been a poverty of modest means. People yearned, however,   1120
To be renowned and to be powerful – that way, they thought
They built their fortunes upon solid ground. But all for naught,

Since as men clawed to the pinnacle of office, all the time
They strewed their path with perils. And at the apex of their
    climb,
Often Envy would blast them like a thunderbolt, to fell
Them with disdain and hurl them in the pit of hateful Hell.
And since, like lightning, Envy loves to singe the summits best,
And anything that raises itself up higher than the rest,
It is far preferable to live in peace and to obey
1130 Than to wish to reign in power and hold whole kingdoms in
    your sway.
Let others wear themselves out all for nothing, sweating blood,
Battling their way along ambition's narrow road
Because their wisdom smacks of others' lips, and they pursue
Things that they only know at second-hand, rather than
    through
Their own senses. For their way of life is just as wrong
Today as it will be tomorrow, and has been all along.

Thus the sad stories of the death of kings: for toppled down
Lay ancient majesty of throne and sceptre; and the crown,
Besmirched with gore, that once had graced the highest head
    of all,
Trod underneath the rabble's feet, lamented its great fall –
1140 For men are eager to trample underfoot what they before
Had held in too much awe. And mankind was reduced once
    more
To chaos, the very bottom of the barrel, as each sought
Power and glory only for himself. Later, some taught
Men to establish a constitution, set magistrates in place,
That people would want to live by laws; because the human
    race,
Weary of leading all their days in violence, bled dry
From constant clashes, were all the more eager to put by
Their own will and submit to the rigid rule of law. Each man
Was ready, out of rage, to avenge himself more fiercely than
Would be allowed under today's impartial laws, and hence
1150 People were sick to death of spending their lives in violence.

It is from this arises the fear of punishment that spoils
Life's gifts, for violence and wrong ensnare all in their toils,
And for the most part, crimes rebound upon their author's head.
Neither is a quiet, retiring life easily led
By one whose hand has shattered society's pact of peace. For
    though
He keep his act from eyes of gods and men, he cannot know
With any certainty that it will stay secret forever.
Indeed, it's said that sometimes in the raving of a fever
Many men blurt out the things that they had buried deep,
While others spill their misdeeds forth by talking in their sleep.   1160

As for religion's origin, that's easy to unfold:
The reverence of the gods that everywhere has taken hold
Of mighty nations – their cities awash with altars – and the ways
Of worship it established, the sacred rites that thrive these days
Among great lands and peoples, and the dread rooted inside
Men even now, that builds the gods new temples far and wide,
And makes folks congregate on holy days – the cause behind
All this is, even way back then, men saw with wakeful mind
The images of gods (and even more so when they dreamed)   1170
Surpassing beautiful and wondrous tall. And since they seemed
To move their limbs, and speak the exalted tones that men
    thought right,
As befit such gleaming aspects and such superhuman might,
Men ascribed sense to them, and granted they would never die,
Since the images refreshed from a perpetual supply,
And their forms were changeless. Especially, men thought beings
    endowed
With such great powers could not by any force be easily bowed,
And that is why they thought them blessed in their happy lot
Beyond all mortals, since the Dread of Death oppressed them
    not.   1180
Also because men saw them in their dreams do many a thing
Marvellous to behold, all without any labouring.

Moreover, men observed the orderly movements of the heavens,
And beheld the cycling of the year with its returning seasons,

But could not fathom how these came about, and lacking
    reasons,
Found an escape by handing these things over to the gods,
And supposed that all things came about by supernatural nods,
And established the seat of the gods up in the quarters of the
    sky,
For it's through the heavens we see night and moon go wheeling
    by;
1190 There, moon, and day and night, and sombre zodiac all turn,
And torches wandering by night, and gliding stars that burn,
And there, the clouds, the sun, the rain, snow, hail storms,
    lightning bolts,
Thunder loud-grumbling its threats and sudden shuddering jolts.

O foolish race of mortals, that gave gods such jobs to do,
Then went and made them fierce with anger into the bargain
    too!
What groans you purchased for yourselves, what grievous injury
For us, what tears you fashioned for the children yet to be!
It is not piety to cover up your head for show,
To bow and scrape before a stone, or stop by as you go
At every altar, flinging yourself upon the ground face down,
1200 Lifting your palms at the gods' shrines, nor piety to drown
Altars in the blood of brutes, nor to chain prayer to prayer;
Rather, to look on all things with a mind that's free from care.

For when we lift our gaze to the vast precincts of the skies
And see the aether spangled with sparkling stars, a wild surmise
About the pathways of the sun and moon enters the mind,
And in the heart already gravely charged with fears, we find
Something begins to rear its head: a new Unease awakes –
Whether that unbounded power of the gods that makes
1210 The bright stars run their various courses govern us on earth.
The mind staggers with doubt when we're faced with the utter
    dearth
Of answers to our questions: did this world once have a birth?
And is there a fixed span of time the world's walls can remain
Before they give way from the unceasing motion and its strain?

Or have those walls been granted by some supernatural force
The strength to last, and slip along forever on Time's course,
Shrugging off immeasurable Age, its crippling might?

Moreover, whose mind does not cringe with superstitious
    fright,
And whose flesh does not creep with awe, when the burnt earth
    shakes   1220
Struck by hair-raising bolts of lightning, and the vast sky quakes
With rumbling thunder? Do not nations tremble, peoples
    quiver?
Do not proud kings, struck by dread of the gods, curl up and
    shiver,
Terrified lest for some despicable crime or cruel command
The heavy day of reckoning is finally at hand?

Or when a tempest rises up, when winds of gale-force sweep
The seas, take the commander and his fleet out on the deep
With all his mighty legions and his elephants of war[31] –
Does he not pray to the gods for peace, and, terrified, implore
The squall to die down, and beseech more favourable winds to
    blow?   1230
But all in vain, since often the violent whirlwind won't let go,
But snatches him up and dashes him upon the shoals of Fate.
So utterly does some invisible force crush man's estate,
Seeming to trample the glorious rods and cruel axes of power[32]
In the dust, as if they were the flimsy playthings of an hour.
Then when the whole earth moves beneath our feet, and cities
    tumble
To the ground, hit hard, or cities badly shaken, threaten to
    crumble,
Is it surprising mortal men are suddenly made humble,
And are ready to believe in the awesome might and wondrous
    force
Of gods, the powers at the rudder of the universe?   1240

But onward: the discovery of metals. Light was first shed
On copper, iron, gold and hefty silver, useful lead,

When fire high on the mountainsides consumed vast forest
        tracts –
Conflagrations sparked by lightning bolts, or by attacks
Of woodland tribes at war who wished to rout their foes with
        fear,
Or else by men who, lured by the fertile earth, wanted to clear
The rich fields and return them to pastureland – or wanted to
        slay
Wild animals and grow rich with pelts plundered from their
        prey
1250 (For men first hunted by means of fire and pitfall, before they
        came
To encircle clearings with nets and to use hounds to flush the
        game).
But whatever it was started the blaze, once the scorching flame,
With an awful roar, had gobbled up the forest, down as deep
As the roots, and baked the earth to a crisp, the burning veins
        would seep,
And trickles of silver and gold, also copper and lead, would
        stream
And pool in the earth's hollows. When these cooled, men saw
        the gleam
Of their glinting colours in the soil, and drawn to what they'd
        found –
The shiny smoothness of the nuggets – pried them from the
        ground,
1260 And saw these bore the shapes of the depressions where they
        lay.
Then this drove home that they could shape the nuggets in this
        way –
Melting them down and pouring them into any mould they
        made.
Indeed, with hammering that they could make as keen a blade
Or sharp an edge as need be, and could fashion themselves tools
To chop down timber, rough-hew logs, to plane beams smooth,
        drill holes,
To punch and pierce. And they were just as likely at the start
To make tools out of silver or gold as of warlike bronze, stalwart

And sturdy; but in vain, because the strength of both would
    yield,   1270
And could not stand up as bronze did to hard work in the field.
For then, since gold was soft and blunted easily, men would
    deem
It useless, but bronze was a metal held in high esteem.
Now the opposite: bronze is held cheap, while gold is prime.
And so the seasons of all things roll with the round of Time:
What once was valuable, at length is held of no account,
While yet the worth of that which *was* despised begins to mount,
And once discovered, day by day's a more sought-after prize,
And flowers in praises, holding the highest honour in men's eyes. 1280

Now as to the discovery of iron and its use,
Memmius, this is something you can easily deduce
For yourself. The weapons of yore were hands, nails, teeth,
    stones, limbs of trees;
Then men made use of fire when they had learned its properties;
Next, the powers of iron and bronze were found out in their
    turn.
But it was first the uses of bronze that people could discern –
Later on they found the strength that lay in iron ore –
Because bronze was more malleable, and lay in greater store.
It was with bronze they ploughed the earth, and fanned the
    winds of war,   1290
And sowed a harvest of wounds, and snatched away livestock
    and field.
The naked and defenceless, before the armed, would readily
    yield.
Then little by little, the iron sword attained the pride of place,
As the sickle, brazen and old-fashioned, fell into disgrace,
Folk started to turn the earth with iron, and when war would
    loom
Between hosts equally matched, the outcome lay in doubtful
    gloom.

Being armed and mounted up on horseback for the fight,
Reining the horse with the left hand, freeing up the right,

Pre-dates harnessing a two-horse chariot to dare
1300 The dangers of war, and *that* pre-dates a car drawn by two pair,
And chariots that men tricked out with scythes for their attacks;
Then dread, snake-handed elephants with turrets on their backs,
Trained by the Carthaginians to bear the wounds of wars
And deal their devastation among the great cohorts of Mars.
Thus one advance after another Dismal Discord spawned,
Woeful for the tribes of Man, and day by day there dawned
Upon the battlefield new horrors in the art of war.

Some also experimented with bulls in battle, and savage boar
Which they set upon their enemies. While other people sent
1310 Brawny lions in the vanguard with a complement
Of armed trainers and cruel handlers able to retain
Control over their animals and keep them on a chain.
But all for naught, since lions would grow frenzied in the fray,
And hot with wholesale slaughter, threw both sides in disarray,
Tossing their terrible heads about, and shaking their dreadful
      manes.
The cavalry could not settle steeds spooked by the roars: the
      reins
Were useless to turn their mounts to face the enemy. In their
      wrath,
Lionesses pounced at anyone who crossed their path.
They launched right at the face of any on-comer with a bound,
1320 Grabbed other riders from behind and dragged them to the
      ground;
Before the men knew what was happening, they were pinned
      beneath
The curving claws, and swooning from the wounds of crushing
      teeth.
As for the bulls employed in battle, these charged their own
      ranks,
And trampled them beneath their hooves, and gored the horses'
      flanks
And gutted their underbellies with their horns, and lowered their
      heads
And tore the earth. The wild boars cut their own allies to shreds

With their sturdy tusks, and maddened by the broken shafts that
    stuck
In their flesh and that were dripping with their own blood, ran
    amok
Among horse and foot alike. You'd see mounts trying to get
    clear
Of the wicked thrusting of the boars' tusks, shy sideways, or rear   1330
And paw the air. But all in vain, because the tusks would slash
Their hamstrings through, and down the steeds would topple
    with a crash,
Sprawled in the dust. Even those beasts the trainers had deemed
    tame
Enough when back at home, they later realized became
Foaming mad, caught in the pandaemonium of war
Among the shouts, the routs, the panic, anarchy and gore.
The trainers tried to call them back, but none of the brutes
    would yield.
Every stripe of beast ran helter-skelter on the field,
As these days elephants will rampage, struck with a botched
    blow,
Wreaking havoc among their friends and laying many low.[33]   1340

If people ever actually tried such things! But as for me,
I find it difficult to think that they could not foresee
The calamitous debacle they thus called down on the head
Of one and all. It's easier to imagine that, instead,
This happened somewhere in the universe, among the many
Worlds constructed in various ways, rather than in any
One particular earth you fancy. But men did not believe
So much they'd win the day this way as cause the foe to grieve;
Outnumbered and unarmed, they were prepared to meet their
    doom.

Then knotted garments came before stuff woven on the loom.   1350
Woven cloth came after iron, alone of all the metals
Able to make the loom's tackle smooth enough[34] – the treadles,
Spindles, shuttles, clacking heddles. Naturally, at the start
Men, not women, spun the wool – for men in every art

Excel the weaker sex in cleverness by far – until
The rugged farmers started calling it a womanish skill.
With weaving deemed a woman's chore, men were happy to
     shirk
The task to women's hands, and take up their own share of
     work –
Backbreaking labour – and by means of that hard work became
1360 Toughened in their hands and hardened in the entire frame.
But Nature herself, Creatress of all things, was first to show
Mankind by her example how to graft and how to sow,
For underneath the spread of trees, the dropped acorns and
     fruits
Sprouted in due season, the ground swarming with new shoots.
It was from observing this, men got the notion to splice young
Scions on the branches of stock, and plant seedlings among
Their fields. And on their blessed little acre of land, they came
To try one method after another, and watched wild fruits grow
     tame
Under the tenderness and coddling of their husbandry.
1370 Day by day, they made the forest pull up stakes and flee
Into the mountains, and yield the lands below for men to till,
That men might have the meadows, cornfields, brooks, on dale
     and hill,
The joyful vineyards – and that hedgerows of olives,
     silvery-green,
Drawing the boundary-lines of plots by running in between,
Spill over valleys, hills and plains – as now you see the farms
Colouring the country with their chequering of charms,
Dotted here and there with orchards of sweet fruit-trees and
With fruitful hedgerows edging round the borders of the land.

Men whistled to imitate the warbling notes of birds a long
1380 Time before they could lift their voices in melodious song
Pleasing to the ears. And zephyrs were the first to show,
By whispering through hollow reeds, the rustics how to blow
Into the hollow stalks of hemlock. From this, by degrees,
Folk taught themselves to play the melancholy melodies

Which pour forth when they press the stops of flutes, a music
    made
Among the untrodden ways of forest grove and glen and glade,
And lonesome refuges of shepherds, out in the open air,
Whiling away the hours. [Thus Time brings each thing, more
    and more
Into our midst, and Reason heaves it up onto Light's shore.][35]
This music eased their minds and pleased them when they'd had
    full measure   1390
Of food and drink, for that's the hour when everything brings
    pleasure.
And often a party of them sprawled on the soft carpet of grass
Beside the riverbank, in the shade of a lofty tree, would pass
The time refreshing their bodies at no great cost. And with more
    reason
When fair weather smiled upon them, and it was the season
That prinked out all the greenery with flowers. That was the
    time
Of joking, and of conversation, and sweet laughter's chime,
For that was when the Rustic Muse was at the height of her
    powers.
Then lusty Merriment instructed folk to go plait flowers
And leaves in garlands, and bedeck the head and shoulders
    round,   1400
And dance galumphing out of time, thumping the hard ground
(Their Mother Earth) with hardened feet. And glee came
    following after:
The clodhoppers would burst out into gales of raucous
    laughter,
For any novelty was all the rage. When men would keep
Watch in the night, music was consolation for lost sleep,
To lift their voices in many ways, and bend a tune, and play
On pipes, with curving lips, as watchmen do to this very day.
Though now they can keep time to many a rhythm, they do not
Take any more enjoyment in the music, not a jot,   1410
Than did that earthborn race of woodland folk in days of yore.

For whatever is at hand, if we have never known before
Anything finer, it gives us chief delight and reigns supreme.
That is, till something better comes along in our esteem
And ruins the earlier thing that's now old-fashioned to our
    mind.
Thus men grew sick of eating acorns, and they left behind
Those beds made out of leaf-litter and grass heaped in a pile.
And clothing made of hides of beasts also went out of style –
Though I imagine, at the start, that finery of that sort
Stirred up such fits of envy that the first person to sport
1420 An animal-skin was ambushed for it and paid with his life,
And then the plotters rent it to shreds amongst themselves in
    strife
And bloodshed, till it was no use to anyone. So of old
With hides of wild beasts, where nowadays purple and gold
Disturb the lives of men with cares, and wear them out with
    war.
And the fault, I think, lies chiefly in ourselves – because before
Animal pelts, the sons of the soil, with nothing on their back,
Were racked with cold, whereas *we* do not suffer from the lack
Of purple raiment worked with massive patterns in gold thread,
So long as we've a peasant cloak to shelter us instead.
1430 Therefore the human race is ever labouring in vain,
And fretting the years away in bootless worries. For it's plain
Man doesn't realize that even *having* has its measure;
There's a point beyond which nothing can increase our real
    pleasure.
And this is what has by degrees dragged Life so far from shore,
And stirred up from the very depths the tidal waves of War.

But it was the shining sun and moon, the watchmen of the
    world,
That as they made their rounds, and quarters of the heavens
    swirled,
Taught Mankind with the cycling of the year, there is a reason,
An order for all things, and for all things, a certain season.

Men were leading lives enclosed in sturdy towers now, 1440
And the land was parcelled into plots, and worked beneath the
    plough,
And the flapping sails of ships already flowered on the deep,
And tribes already had leagues and alliances to keep
When poets first began recording human deeds in song,
And the alphabet had still not been in use for very long.
And that is why we cannot peer into the past far back,
Unless Reason is somehow able to set us on the track.

Seamanship and agriculture, law, fortifications,
Weaponry, clothing, roads, and all these sorts of innovations,
The prizes of life and absolutely everything that's fine – 1450
Poetry, painting, statues cunningly wrought and made to shine –
By trial and error, and probing, restless intellect, day by day,
Step by step, these skills were taught to Man, feeling his way;
So incrementally Time brings all things within our sight,
And Reason lifts them up into the boundaries of light.
For men saw one thing after another clearly in their hearts
Till they ascended to the very summit of the arts.

BOOK VI
WEATHER AND THE EARTH

Athens the Illustrious was first in bygone years
To teach the art of cultivating the grain's fruited ears
To ailing mortals;[1] it was she who made life new again,
And also she who instituted laws to govern men.[2]
And she was first to bless life with sweet solace, giving birth
To a genius of a man,[3] a man with mind of greatest worth,
Who poured from his truth-telling lips all things of heaven and
    earth.
And though his life was snuffed out, yet his glory shall not die –
It spread of old with his good news, and now reaches the sky!

For when he realized that most things people could demand
10 As necessary for their livelihood lay ready to hand,
That they'd set life, as far as possible, on solid ground,
And that, awash in wealth and privilege, powerful and
    renowned,
And blessed by the good name of sons, nevertheless, apart
In the privacy of their own homes, each was worried sick at
    heart –
Against their will, these worries plagued their life, with no
    relenting,
And made them burst out bitterly with cursing and lamenting –
He understood it was the vessel itself that was to blame,
And that by means of its internal flaw, all things that came
Inside it from without, however sweet at first, went sour;
20 Partly, he saw, since it had holes and leaked, and by no power
Could anything fill it; partly since when anything was placed
Inside the vessel, it was 'corrupted' with an evil taste.

And thus with his truth-telling words he washed the heart all
   clear,
And set a limit to desire and an end to fear,
And showed what was the highest good, towards which we all
   strain,
And pointed out the route, that we, unswerving, might attain
The strait and narrow path; and showed what evils and what
   cares
Worry man at every turn and darken his affairs;
What ways, by fluke or force of Nature, these take wing; and
   more –   30
He showed how each type should be met, by sorties through
   which door,
And demonstrated that mankind in vain, for the most part,
Set the gloomy sea of troubles churning in the heart.
For just as children shiver and dread all things in black of night,
So *we* sometimes are terrified of things in broad daylight
That are no more to be feared than what sets tots to trembling
In the dark, the bugbears that they fancy are about to spring.
The fear and shadows of the mind must be scattered away,
Therefore, not by the sun's rays nor the shining spears of day,   40
But by the look of Nature and her law. Thus all the more
Reason to take the thread up where I had left off before.

And since I've shown the precincts of the world are doomed to
   die
And that there is a birthday for the substance of the sky,
Since I've unravelled events that take place there, and have made
   plain
The things that must take place, listen to what topics remain,
Once I mount the glorious car [of Poesy

                       to explain]
The winds, and how their fury is stirred up, and how it's eased,
And all is calm again once the winds' anger is appeased,
And other things men see on earth or happening at a height,   50
That often keep their wits on tenterhooks, distraught with
   fright,

Phenomena which cow their spirits and which bow them low
Upon the earth with awe of gods. For since men do not know
What causes lie behind events, they are left to explain
These as the dominion of the gods, submitting to their reign.
[Because they cannot see causes of things, then they opine
That things are done by means of powers mighty and divine.]⁴
For even those who are well versed in how the gods live free
From care, once they start wondering how these things come to
    be –
60 Especially those events that they see happening overhead
In the shores of heaven – backslide into superstitious dread:
They take up harsh taskmasters whom the wretches think are
    All-
Powerful, because they don't know what is possible
And what is not, and that the power of every thing must keep
Within determined limits, with the boundary stone set deep.
How far astray they blunder when their reasoning is blind!

Unless you spew out these wrong-headed notions from your
    mind
And banish from yourself all thoughts unworthy of the gods
And alien to their peace, their power will often be at odds
70 With you and do you harm, that holy power that you slight;
Not since their sublimity is outraged, and they thirst to smite
In wrathful, fierce revenge, but since you'll have it in your head
That tranquil beings, living in unruffled calm, instead
Seethe with great waves of wrath. You won't be able to draw
    near
The temples of the immortals with a heart that's free from fear.
As for those images that wend their way to minds of men
From their holy bodies, and whose stamp declares their origin
From forms divine, you won't be in a fit state to receive
Them with a peaceful spirit. And it's easy to perceive
Right away the kind of life this leads to! Thus, although
I have already covered a lot of ground, yet even so,
80 To save us by most true philosophy from such a curse,
There are yet many things I must adorn with polished verse:

We must understand the look and working of the skies,
And I must sing how storms and lightning's dazzling flash arise,
What causes them and what they wreak – lest you, out of your
    wits,
Section up the sky for augury in different bits,[5]
And shudder to behold from what direction lightning flits
And into what quarter of the sky it travels, in what way
It slithers into a closed space, and holds it in its sway,
And after, darts outside again. [For since men cannot see   90
The causes of things, they reckon it is by gods' agency.]

O Crafty Muse, Calliope – man's respite, gods' delight –
Now show me to the finish line that has been chalked in white,
Guide me, winning near the goal, so I may gain the crown
Of victory in the race, with admiration and renown!

First, the reason that the thunder shakes the azure sky
Is that the scudding clouds crash into one another high
Up in the aether when the winds are warring, for no sound
Comes out of a clear patch of sky; but wherever clouds are
    found
Mustered in thicker ranks, it's from that corner that the
    rumbling   100
Thunder's usually more apt to mutter mighty grumbling.
Besides, the stuff of clouds cannot be dense as wood or rock,
Nor can they be as wispy as the mist or wafting smoke,
Because they'd either sink like stones, sagging beneath dead
    weight,
Or else like smoke, failing to cohere, would dissipate,
Unable to contain the snow or keep the hailstones pent.

Clouds also make a sound above the unfurled firmament,
As canvas, stretched across a spacious theatre like a tent,[6]
Snapping and cracking between the guys and poles, will
    sometimes tear   110
And writhe about, tossed by the wanton currents of the air,
Making the kind of sound a ripping piece of paper makes
(A kind of sound that you can hear in thunder when it breaks),

Or when a buffeting breeze slaps at a sheet hung out to dry,
Or sheets of paper, snapped up by the wind, go flapping by.[7]

Sometimes it also happens clouds do not so much collide
Head on but, rather, pass each other, scraping side on side,
Their bodies dragging in opposite directions as they slide,
Making a continuous crackle that grates upon the ear
120 Until they've worked themselves out of that tight squeeze and
     got clear.

Here's another way that often a booming clap of thunder
Seems to shake and shiver everything and rend asunder,
Suddenly, the mighty ramparts of the compassing world:
When all at once a hurricane of blasting winds has hurled
Together and twists itself into the clouds and starts to race
Round and round, holed up inside, it hollows out a space
Within the cloud, so that on all sides, as the whirlwinds spin,
The substance of the cloud is squeezed to make a thickened
     skin;
Then, after, when the ferocious onslaught of the winds has
     worn
The cloud's walls thin, there's an ear-shattering crack as it is
     torn.
130 And it's no wonder, seeing that even small balloons filled up
With air will make a whopping *bang!* that startles when they
     pop.

This is another possible cause of thunder: as winds blow
Through clouds, they make a racket. We often see that as clouds
     go
Along, that they are curiously branched and deckled, and we
     know
That when, through a dense wood, the gusts of a northwester
     smash,
The foliage gives a roar and limbs and branches creak and crash.
It also sometimes happens a strong wind on its swift course
Rips through a cloud, shattering it with the brunt of head-on
     force,

For it is plain what wind can do up in the atmosphere
When we see what it does on *earth*; it's not as strong down here,   140
Yet it pries lofty trees, deep roots and all, out of the ground.
And there are waves up in the clouds that, breaking, give a
    sound,
A low rumble, just as deep-running rivers make a roar,
Or mighty seas make when the breakers crash upon the shore.

Thunder also happens when a burning lightning bolt
Zaps from cloud to cloud. And when the cloud that takes the
    jolt
Of fire by chance is soaking wet, it makes a terrific hiss,
Quenching it at once, as iron white-hot from the furnace
Sizzles plunged down into ice-cold water standing by.
But if the cloud that takes the bolt instead is rather *dry*,   150
It kindles instantly and burns up with a noisy whoosh,
As fires whipped by gales of winds across a hilltop lush
With laurels sweep through and consume them in a violent
    rush –
And there is nothing makes a din so awful when flames swallow
Them up as do the crackling Delphic laurels of Apollo!
And last, a cracking of ice, tumble of hail, create a loud
Shattering clatter oftentimes high up in heaps of cloud
When mountains of thunder-heads are crushed together by a
    gale
Into a narrow space and crumble, crunched up with the hail.

Lightning also flashes when the clouds, colliding, knock   160
Out a number of seeds of fire, as when you strike a rock
On flint, or flint on iron, for in that case too, light leaps out
And sends the glittering sparks of fire scattering about.

But why is it we hear the thunder *after* the flash appears
To our eyes? Because the particles that travel to our ears
Always take longer reaching us than those that reach the eyes
And trigger sight. Here's an example you can recognize:
If you see someone far off with a double-headed axe
Felling a massive tree, you see the strokes before the thwacks

170 Reach your ears. Thus also we discern the lightning's flame
      Before we hear the thunder, though the cause of both's the same,
      And both arose at the same time, born of a single crash.

      Another way clouds light the landscape with their flickering
            flash,
      And the storm glimmers with lightning, is that a wind will spin
            and spin
      Once it has got inside a cloud, and hollows it within,
      Pressing its substance into a thick wall, as I have taught,
      And grows hot with the swirling, just as anything grows hot
      And kindles with movement: take how leaden bullets when
            they're shot
      Over a long distance actually grow molten from turning.
180 Therefore, when a gloomy cloud's rent open by a burning
      Wind, the sudden force of the explosion is to blame
      For scattering the seeds of fire that make the flashing flame,
      And the rumble follows, taking a longer time to reach the ear
      Than it takes the flash to reach our eyes and lightning to appear.

      This is indeed what happens when at the same time clouds are
            dense
      And heaped on one another in piles towering and immense.
      In case you are deceived by their appearance from *below*,
      Where we can see how wide they spread, but not how high they
            grow,
190 Look closely when wind sails the mountainous clouds across the
            sky
      Or when you see clouds on the lofty mountaintops stacked high
      And pressing down upon them, cloud on cloud piled in a heap,
      At anchor while the winds on every side lie sunk in sleep,
      Then you will understand how massive their peaks are, and your
            eyes
      Will make out deep-domed 'caverns' carved in cliffs hung in the
            skies.
      The winds fill up these caverns when a squall begins to rise,
      And trapped within the clouds they give a terrible roar of rage,
      Snarling like wild animals locked up inside a cage;[8]

Pacing throughout the clouds, now here, now there, they roar
    and growl,   200
Seeking an avenue of exit, round and round they prowl,
And picking up the seeds of fire from the clouds as they spin,
And amassing numbers of them, roll the flame about within
The hollow furnaces, and cracking them open, flicker and glow.

Another reason why the golden forks of fire that flow
So nimbly dive to the earth is that the clouds themselves contain
A multitude of seeds of fire, since when they hold no rain
They are most luminous of colour, brilliant and bright –
Indeed they must absorb many such seeds from the sun's light –
So it is only natural they blush and shower fire.   210
And therefore when a driving wind has rammed the clouds
    together,
And crammed them into a close space, clouds squeeze out and
    let flow
The seeds that make the flickering of fire flash and glow.
Lightning also happens when the clouds thin in the sky;
For when the wind lightly disperses clouds as they go by
And ravels them, the seeds of fire willy-nilly tumble,
And lightning flashes forth without the sickening shock and
    rumble.

What's more, the nature of thunderbolts is easy enough to find
From scorch-marks and the branded burns of heat they leave
    behind,   220
And from the oppressive fumes and whiff of brimstone that
    remain –
For these are Fire's hallmarks, not the sign of Wind or Rain.
Furthermore, they often kindle housetops and then – poof!
The quick blaze takes dominion of all things beneath the roof.

You may be sure this fire is the most fine-spun fire of all,
And Nature formed it out of particles so quick and small
That there is nothing whatsoever can obstruct its flame.
For a powerful thunderbolt passes through closed doors just the
    same

As noise or voices, goes through stone, through bronze, will
    liquefy
230 Bronze and gold in a trice. It suddenly drains the wine jars dry,
Careful not to break the jars themselves. No doubt the way
It does this is the lightning's heat as it comes near the clay,
Easily loosens the pottery's fabric, making it rarefied,
So that it's able to slink within the jar, and once inside,
Quickly frees and scatters abroad the wine's atoms – a feat
The sun cannot perform in a whole lifetime with his heat,
Mighty as he is with his flashing fire. Thus how much faster
Thunderbolts must be, and how much more their force is
    master!

Now then, what causes thunderbolts and why they have such
    power
240 That they are able with a blow to blast apart a tower,
To raze houses, wrenching the beams and rafters out, to lay
Low the monuments of men or shudder them, to slay
People, and to slaughter livestock wholesale – by what might
They're able to wreak these kinds of havoc, I shall bring to
    light
And shall not stall you any more with promises.

                                       It's clear
They come from thick, high-builded clouds, because they never
    appear
Out of a clear blue sky, nor when the cloud cover is light.
Plain facts you can observe will demonstrate that this is right –
Because when lightning-bolts *do* strike, they come from such a
    doom
250 Of clouds amassing in the sky, that we imagine gloom
From every direction has come swarming out of Hell
To throng the towering dome of Heaven, there is such a fell
Midnight of clouds that gathers in the firmament overhead
Out of which there leer the countenances of Black Dread
When the storm gears up to hurl its thunderbolts. And
    furthermore,
It often happens that a murky cloud far out from shore,

Like a river of pitch rained from the sky, and crammed with
    shadows, falls
On the waves, dragging behind it a black hurricane great with
    squalls
And thunderbolts, the cloud itself near bursting with a welter   260
Of wind and fire, so even on dry land men run for shelter
Shaking with fear. Thus we should understand such tempests
    surge
To towering heights above our heads, since clouds could not
    submerge
The earth with so much dark unless a host of them piled one
Upon the other reached up high enough to block the sun,
Nor could such cataracts of rain come pounding down to brim
The rivers into flood, and make the meadowlands aswim,
If the aether were not stacked up high with clouds. And it
    transpires
In such a case that all the heavens teem with winds and fires,
And everywhere the roars and flashes that such weather breeds.   270

As I have shown above, the hollowed clouds have many seeds
Of heat, and must absorb a number from the sun's warm rays.
Once a wind has chanced to gather these into one place,
It squeezes many seeds out, and in the process is combined
With this fire. There the whirling current swirls and swirls,
    confined
In narrow quarters, and, within that fiery furnace, whets
Its thunderbolt. Wind catches fire two ways: it either gets   ·
Hot from its own movement, or from fire's contact. When   280
The wind is white-hot and a burst of fire has charged it, then
The thunderbolt, now ripened as it were, suddenly tears
The cloud to shreds, and sends a bright blaze hurtling out that
    flares,
Lighting all the landscape with its flash. Next in its wake
Comes an ear-splitting crack, so that the heavens seem to break
Open and come tumbling on our heads. And then the ground
Shudders beneath our feet, and rumblings go racing round
The towering skies, for almost all the storm reels from the force,
And gives out groans. Following the shock, a cloudburst pours   290

In such a spate it seems the whole of heaven melts to rain,
And the Great Flood[9] threatens to engulf the earth again,
The broken cloud and gale of wind unleash so great a flow
Of water when the sound comes crashing from the fiery blow.
Or sometimes an *external* blast of wind swoops with a jolt
Upon a cloud that's gravid with a full-term thunderbolt,
And as it rips the cloud, the burning whirlwind's suddenly flung
Down – that thing we call a thunderbolt in our mother tongue –
Although, depending on the angle of the blast's attack,
300 Bolts can fly in any direction.

                                        Sometimes, despite a lack
Of fire initially, a wind can *catch* fire as it blows
On its long journey, shedding bulky atoms as it goes
That drag against the air, while other atoms small in size
It rakes up out of the air and carries off, and as it flies,
Mixing the fiery atoms with itself, bursts into flame.
The way lead bullets grow hot in their course is much the same –
Shedding a layer of cold, stiff atoms they catch fire mid-flight.

It may be that the force of the crash itself sets things alight
310 When a chilly blast of wind that has no fire strikes a blow;
No doubt, since in the violent strike the seeds of heat will flow
Together from the object that receives the sudden shock,
And from the wind itself – just as, when we strike flinty rock
With iron, flame leaps from the blow, nor does it thwart the
        spark
Of fiery atoms that the stuff of iron is cold and stark.
So anything may kindle from a thunderbolt in turn,
If it be made of stuff combustible and fit to burn.
Nor can a wind that has been hurled from on high with such
        force
320 Be absolutely cold. Even if early on its course
It has not kindled, by the time that it arrives it will
Be mingled with the seeds of heat, be hot instead of chill.

The nimbleness of thunderbolts, the violence of their blast,
And how they are unleashed and plummet hurtling down so fast,

Is first of all because there is a force up in the rack
Of clouds that's always restless, collecting itself for the attack
With mounting energy, and when a cloud cannot hold back
The built-up power any more, it shoots the thunderbolts
With amazing force like missiles hurled from massive catapults.

Add to this, a thunderbolt is made of atoms small   330
And smooth, so that it's difficult to block its way at all,
Since it can slip between and slide through pores and
       passageways,
Nor does it balk at many hindrances so it delays,
And that is why it zips along at such a spanking pace.
Then also, while by nature all weights have a downward pull,
Augment this with a push and the velocity will double:
The downward urge grows heavier, and so with greater speed
And force the bolt dashes apart whatever may impede
Its progress, hewing to its course. The thunderbolt, indeed,
Because it travels on a journey of such distance, grows   340
Ever speedier, the speed increasing as it goes,
Building up its power and steeling the impact of its force.
Its velocity directs the seeds to go straight on their course
Together on a single path. And maybe on its flight
The bolt drags certain bodies out of thin air that ignite
The rushing with their blows.

                                   The reason a thunderbolt can go
Through many things while leaving them intact is fires flow
Like liquid through the pores in things; while other objects
       shatter
Because the lightning's atoms strike the junctures of their
       matter,   350
The interconnections of their fabric. Or take gold or bronze –
A bolt of lightning easily will melt these down at once
Because its force consists of particles most fine and small,
And slippery elements, so that it is able to crawl
Easily inside, and of a sudden, once it slinks
Within, it unties all the knots and loosens all the links.

The hall of heaven hung with glowing stars most often quakes
With thunder in the autumn; it is then the whole earth shakes,
And also in the flowering of spring. For in the dead
360 Of winter there's a dearth of fires, while there's a lack, instead,
Of winds in summer's heat, and then the clouds are rather thin.
Therefore the temperature of the skies must fall somewhere
        within
Those two to mix conditions that are right for thunderweather;
For when, in the year's straits, the two cross-currents churn
        together,
They mingle hot and cold (because the clouds need both to
        form
The bolts of lightning that they forge up in a thunderstorm),
So that things are at odds, and the air furiously raves
In a mighty clash of winds and fires, and surges in wild waves.
For when warmth is beginning, cold is coming to its close
And this is the season we call *spring*, when unlikes must
        oppose,
370 And mix and throw things into uproar. When the heat has
        rolled
Round to its ending, it is mingled with the start of cold,
And here we have the season that we call the autumn-tide,
Again where bitter winter and the summertime collide.
And this is why we call these seasons the straits of the year,
It is no wonder this is when most thunderbolts appear,
And when the seething storms are stirred up in the
        atmosphere,
Since the sky is all at war, as both sides wage a close
        campaign –
One side with fire, the other with joined ranks of wind and
        rain.

This is how to understand the thunderbolt, its nature
380 And by what power it does its deeds – and not by poring over
The mumbo-jumbo furled up in Etruscan scrolls[10] to strain
To grasp signs of the hidden purpose of the gods, in vain,
By noting of the burning bolt which corner of the skies
It starts from, and towards which quarter it then turns and flies,

And how it slithers into a closed space, and in what way
It flees again once it has held the whole place in its sway,
And how a bolt out of the heavens can pollute a place.[11]

But if Jove and the other deities[12] shake the shining face
Of the sky with a bone-jolting crash, and if the gods can throw
Fire whatever direction they like, then why don't they lay low
Scoundrels who flaunt a heinous crime, so that, pierced through
    the breast,   390
They breathe out flames and serve as a dire warning to the rest?
Why is it someone else instead, all innocent of blame,
No crime upon his hands, is suddenly engulfed in flame
And snatched away by a fiery whirlwind swooping from the sky?

And as for wasting good throws on deserted places – why?
To keep in practice and firm up their biceps? Have they found
Their Father does not mind his bolts are blunted on the ground?
Why does he stand for it, instead of saving them for his foes?
And why again does Jupiter never hurl one of his blows   400
To the earth and crack his thunder when the sky is blue and
    clear?
Does he wait for clouds to mount, so he can climb inside and
    steer
The bolt on its trajectory while he is standing near?
Why does he smite the sea? What can his purpose be, what
    whim?
What have the swimming plains of whitecaps ever done to *him*?

Besides, if he intends we should be wary of his blast,
Why does he hesitate to let us see it as it's cast?
And if he wants to smite us unawares, why does he blunder
And give us warning to take cover by the threat of thunder?
Why first gloom's adumbrations, rumours rumbling, I wonder?   410

And how explain when he fires many directions at one go?
Or do you dare to say he never fires more than one blow
At once, that this is something never done – when it is plain
This *often* happens, and *must* happen, seeing that the rain

Pours down in many regions simultaneously. Likewise,
Many bolts of lightning fall together from the skies.

And last, why does he blast the sacred shrines of gods asunder
And wreck his *own* renowned abodes with brutal bolts of
      thunder?
Why does he smash the well-wrought images of gods and maim
420   His own statues, disfiguring their splendour with his aim?
Why as a rule is it the higher places that he seeks;
Why do we find his scorch-marks mostly on the mountain
      peaks?

It's easy from the earlier explanations to see why
What Greeks call 'blowers' – waterspouts – descend out of the
      sky
Onto the water. For, at times, a kind of column falls
Out of the sky and touches the sea. And whipped up by the
      squalls
Of heavy, hissing blasts, the sea around it seethes and brawls.
And any bark that's caught out in the hurricane is tossed,
430   Pitching and yawing on the waves, in danger of being lost.

Sometimes this happens when a stirred-up force of wind at first
Attempts to rend a cloud, but when it cannot make it burst
Presses it down instead, so that it drops upon the seas
Like a column let down from the heavens, as though, by degrees,
Something inside the cloud reached towards the water from
      above –
As if a fist on outstretched arm pushed down a cloudy glove.
And when the force rips open the cloud at last and it bursts free
On the waves, it unleashes an amazing boiling of the sea,
For as it's lowering, the whirlwind twists, and it drags down
The pliant substance of the cloud along with it, but soon
440   As it thrusts the pregnant cloud on the face of the waters, in a
      flash
It plunges itself in the waves, and roils them up with a
      monstrous crash.

Or at times a whirling wind will wrap itself inside a shroud
Of mist by scraping together from the air the seeds of cloud
And imitates a 'blower' that's been lowered from the sky.
And when it gusts in off the sea and touches down on dry
Land, it disperses itself and vomits forth an enormous force
Of funnel clouds. But this is rare, and as a matter of course
The mountains block our view on land. So it's out on the deep
We tend to see them, against the whole sky's panoramic
    sweep.   450

Now, clouds are formed when in the upper reaches of the sky
Many floating bodies come together suddenly,
Rather rough in texture, so that they can cling, despite
The fact that bonds among the particles of cloud are slight.
The seeds combine together to make cloudlings first, and these
Catch onto one another, accumulating by degrees,
And driven by the wind in racks, they gather in a swarm,
Until the point when they build up into a raging storm.

It also happens that the higher up the mountains crowd
Against the sky, the more their peaks continuously cloud   460
With a thick scrim of tawny mist, because when clouds begin
To form – before the eye can make them out, they are so
    thin –
The wind harries and herds them up against the summits. Here
At length they flock together thickly enough that they appear,
While seeming at the same time to rise from the mountain
    peaks
Into the aether. Indeed, experience itself bespeaks
The windiness of upper altitudes; for when we scale
Towering mountain steeps, continuous blasts of air assail
Our senses.

           In addition, the fact that laundry hung up near   470
The seashore becomes damp with clinging moisture makes it
    clear
That Nature is ever lifting a host of particles from the sea,
Making it probable that clouds swell from the quantity

Of particles rising from the briny tossing of the tides,
The moisture in either case entirely akin.

                                        Besides,
We notice that from all the rivers and from the very earth
Mists and vapours rise, exhaling from the land like breath.
And in this manner, steams are borne aloft, and steep the sky
480 In murkiness, and gathering by degrees, they resupply
The lofty clouds. For also the seething of the starry aether
Pushes from above and squeezes the vapours close together
So that they weave a tapestry of cloud beneath the blue.

The bodies that make clouds and scudding racks, it's also true,
Can enter our heavens from beyond. For I have shown the sum
Of atoms is innumerable, and the Deep from whence they come
Is bottomless, and I have shown how nimbly and how fast
They speed, how in a trice they regularly cross such vast
Distances as boggle the mind. It's therefore no surprise
490 It doesn't take much time for gloom and storm hung from the
         skies
To bury underneath huge thunderheads both sea and land,
Since on all sides, through every corridor of aether and
Through all the breathing pores and passageways of the whole
         world round,
Entrances and exits for the elements abound.

Now pay attention, and I'll show how moisture of the rain
Is collected in the clouds above and is sent down again
To the earth in showers. First I'll prove that many seeds of water
Arise with the clouds themselves from all things. And both grow
         together,
500 The clouds and whatever moisture the clouds hold. It is the
         same
With us, the flesh grows with the blood, or anything the frame
Contains by way of moisture: sweat, et cetera. Like wool
Fleeces hung out on the line, clouds also often pull
Large amounts of water from the sea when winds convey
Them over the vasty deep. And water in the selfsame way

Is lifted to the clouds from all the rivers. When enough
Water-seeds accrue in various ways and build and stuff
The clouds to bursting from every side, the saturated clouds
Strive to release their moisture in a twofold way: wind crowds   510
The clouds together with its force, and heaps them in a tower,
And the pressure of their own mass bearing down wrings out a
     shower;
Whereas when clouds are loosened by the winds or come
     undone,
Scattered thin when they are beaten by the heat of the sun,
They give off drops of rain and drip, just as candles grow lax
When held above a burning flame, and run with melting wax.

Two factors are at work when violent downpours come to pass:
The clouds are violently pressed by winds and squeezed by their
     own mass,
While steady rains that don't let up and last for quite a while
Happen when many water-seeds arise and cloud-racks pile   520
On sopping clouds, gathered from every quarter, and the
     terrain,
Steaming everywhere, exhales the moisture up again.
And it is at such times, when the sun shines through the murky
     rack
With its rays, and raindrops of the clouds it strikes against gleam
     back,
That the colours of the rainbow glow against the tempest's
     black.

And all the other things created overhead which grow
Condensing in the clouds, yes every one of them – the snow,
The winds, the hail, the chilling frost, ice with its power to
     stay
The waters, Mighty Hardener that makes rivers delay,   530
Reining them back along their headlong course at every turn –
How and why these things arise is easy to discern
For all their variety – the mind discovers and it sees –
Once you have grasped their elements' inherent properties.

Now come and learn what is the cause of earthquakes. You
    should know,
First of all, that as the earth's above, she is below,
With draughty caverns all around, and in her lap she keeps
Many lakes and pools and many broken crags and steeps,
540 And we must picture many secret streams beneath her crust
Tossing stones submerged in their violent churning, for she
    must
Be everywhere of one consistent nature, as facts show.

And therefore, since there are such landscapes located below
Yet attached to the surface, the upper world shakes from the
    shock,
When underneath Time undermines the vast caverns and rock
Collapses; indeed, whole mountains tumble down, and in a
    flash
Tremors ripple far and wide out from the massive crash –
And rightly so, since houses by the shoulder of the road
Shudder when wagons trundle by, even if they're light of load;
550 No less so does the tumbrel buck and jostle when we feel
It hit a rut that jolts the iron rim on either wheel.

It also happens that over time a monstrous chunk will break
Loose from the earth and tumble headlong into a vast, deep
    lake,
So that the churning wave causes the earth to tremble and
    quake,
Just as sometimes a vase of water can't quit shaking till
The liquid that is sloshing back and forth inside grows still.

Besides, sometimes a wind collects there in the earth's recesses
And, surging up from one direction, thrusts against and presses
Into the dizzying caverns with a mighty blast, and throws
560 The earth off-balance so it lurches where the wind-blast blows.
And then the houses built upon the surface, the higher each
Is erected, and the further towards the heavens that they reach,
The more they lean to one side and they dangerously sway,
And the beams jut out, at any moment ready to give way.

Yet people are afraid to think the great world has a date
Of doom in store for it, and a demise that lies in wait,
Even when they witness such a bulk of earth tip over!
But if the winds did not inhale again, there is no power
Could keep the world from plunging towards its ruin and rein it
    back.

But as the winds alternately build up and then grow slack,   570
Now rallying their forces, as it were, for the attack,
Now beaten, making their retreat again – it is because
Of this earth *threatens* to collapse more often than it *does*.
It lurches forward and springs back. For having tottered, then
The earth regains its place and gets its balance back again.
This therefore is the way all buildings sway – the top part
    sways
More than the middle part, the middle swings more than the
    base,
And the bottom scarcely budges.

                    The same great shaking comes about
When a wind or powerful blast of air, arising from without,
Or from the earth itself, suddenly dives into the bowels   580
Of the earth, and whirling round in the towering caves at first it
    howls,
Then having built its force up, storms its way out to escape,
Splitting the earth from deep below and making it yawn and
    gape.
This is what struck Sidon in Syria, and in the Peloponnese,
What happened to the town of Aegium,[13] when a release
Of air and the temblor that it triggered made those cities fall.
Strong quakes inland have levelled many a fortifying wall,
And many a coastal town's sunk to the bottom of the waves
With all of its inhabitants consigned to watery graves.   590

But if the wind does *not* burst out, and if instead the force
Of raging air disperses as a shudder through the pores
That riddle the earth, it triggers tremors thus: as when a chill
Seeping into our limbs will make them shake against our will

And shiver. So in cities, men are racked with double fears –
Worried that buildings overhead might crash about their ears;
Likewise in terror of the caverns that are lying under,
Lest suddenly the earth unravel, or she split asunder,
Cracking open into a wide chasm, and she yaw,
600 Seeking with her own debris to cram her gaping maw.

Concerning earth and sky, men may insist however they please
That both are indestructible and hold an eternal lease
On life; from time to time, however, danger itself, when near
At hand and threatening from some quarter, pricks them with
        the fear
That suddenly the bottomless pit could open up and swallow
The earth beneath their feet, and that the Sum of Things would
        follow,
Tumbling down, foundered to its foundations – all things hurled
Together in the rubble and the ruin of the world.

Men wonder,[14] first, the reason nature does not overfill
The sea, when such a flow of waters and all rivers spill
610 From every side into its basin. Add the shifting rain
And scudding storms that drench and drown both dry land and
        the main,
And add the sea's deep sources, yet all this will scarcely count
As a single drop compared with all the ocean's vast amount;
Thus it is not so wondrous that the sea does not increase.

Besides, sun with his heat evaporates great quantities,
For indeed we see that laundry, sopping wet, dries in the sun
From the burning of his rays. And yet the seas are legion
As we can see, and stretch out in a vast expanse. So think:
620 However small a quantity of water the sun may drink
From any given spot, nevertheless the sun must drain
Huge amounts from the broad surface of the sea. Again,
Winds also pick up quantities of water when they sweep,
Skimming off a large share from the surface of the deep.
Indeed we often see the soft mud on a road, within
The space of a single night, dry out and form a scaly skin.

And further, I have demonstrated how clouds lift away
Moisture absorbed from the broad face of the ocean, and how
    they
Disperse it over the whole globe, sprinkling here and there,
When it rains upon the land and winds blow cloud-rack through
    the air.   630

Lastly, since the body of the earth is full of pores,
And wedded to the sea that everywhere engirds her shores,
It has to be that, just as water flows into the deep
From land, so moisture from the salty sea likewise must seep
Into the land, filtering the brine out as it goes,
So only the pure water remains, and this supply then flows
To the sources of the rivers, and runs overland in sweet
Columns, trooping down the paths cut by its liquid feet.

Now I'll set forth the reason that from time to time fires
    breathe
Out of the jaws of Aetna, how they swirl around and seethe.   640
For it was with no middling mayhem that the fiery storm
Lorded it over the fields of Sicily, when in alarm
Her neighbours lifted their eyes to her, and saw in all the parts
Of heaven, flashing smoke, so that a horror thrilled their hearts
That the devastation they were witnessing as it unfurled
Was Nature toiling towards the overthrowing of the world.

You must look into subjects such as these both far and deep
And consider them from every angle and with broadest sweep.
Keeping in mind how fathomless is the Sum of Things, the All,
You'll see a single sky is insignificantly small,
A tinier portion of the universe, for what it's worth,   650
Than is one human being in relation to the earth.
But if you keep this basic tenet constantly in view,
And see it clearly, many things will cease to puzzle you.

Do we find it *mysterious* if somebody has caught
A fever that goes burning through their limbs and makes them
    hot,

Or if some other ailment racks their body? In a trice
The foot swells up, sharp pangs throb through the teeth, even
     the eyes
660 Are pierced right through with pain, the Sacred Fire[15] begins to
     burn
Creeping over the body and seizing every part in turn,
Slithering through the limbs – doubtlessly since there is seed
Of myriad things in the world, and our earth and the heavens
     breed
Disease enough to bring forth pestilences by the score.
Likewise we must grasp that, out of the Infinite, a store
Of every kind of atom supplies the entire earth and sky,
Enough to shock the earth and start it quaking suddenly,
Or make a fierce tornado tear across the land and sea,
Or the fires of Aetna overflow, or the sky go up in flame –
670 For the regions of the heavens catch fire also – it's the same
For thunderstorms that pour down with a very heavy spate:
Atoms of water amassed at random from the void.

                                        'But wait –
The eruption's blaze is too immense for such an explanation!'
Very well. And any stream will seem to be, to one
Who's never seen a larger, the greatest of rivers. And a tree,
A man, will seem enormous – indeed, anything we see
We shall imagine is the largest specimen of its kind
If it's the largest we've laid eyes on. Still, if you combined
All these things together with the earth, the sky, the sea,
They still would be as naught compared with the totality
680 Of the universe.

                    But now I shall explain how from the vast
Crucibles of Aetna this fire sparks and breathes its blast.
First of all, the mountain's hollow in its fundament,
And stands supported in the main by arching vaults of flint.
Again, the wind and air blow through these tunnels
     underground –
The wind arising when the air's stirred up and blown around.

And when the wind has heated up, and raging round the rocks
And earth, has made whatever it touches scorching hot, it
    knocks
The rapid flames of hot fire out of them, and then it draws
Itself upwards and shoots sky-high straight through the
    mountain's jaws.
Thus far and wide it casts the blaze and scatters ash, and
    smoulders   690
With thick and gloomy smoke, and vomits up enormous
    boulders –
You cannot doubt the wind's strength, seeing the force with
    which it shoots!

The sea breaks its waves on a great stretch of the mountain's
    roots
And swallows its surge back again, while tunnels from the sea
Reach underground right to the mountain's towering throat. So
    we
Must understand that water penetrates it from the tide
Of the open sea, and hurls itself and a jumble of things inside,
And that this influx makes the flame go blasting upwards and
Spring leaping out, and heaves up rocks and kicks up clouds of
    sand;   700
Indeed, Sicilians call the summit a 'krater',[16] to denote
'Mixing bowl' – the part we call the mountain's mouth or
    throat.

There also are some things for which *one* reason will not do,
Which require many explanations, though only one is true.
Imagine, for instance, that you spied a body robbed of breath
Lying in the distance – you might weigh all causes of death
In hopes of hitting upon the right one. For you cannot know
With certainty if it was knife or chill that laid him low,
Or disease, or maybe poison. These are the kinds of things we
    name   710
Knowing something of this sort is bound to be to blame.
And we must deal with many other phenomena the same.

Egypt's Nile is the only river in the world that spates
And floods the fields in summertime. It often irrigates
Egypt in the midst of heat waves, maybe since the mouth
Of the Nile is opposite the yearly winds[17] then gusting south
And blowing against the current, and these stem the flow, and
     rout
The waves, and make the river stand and flood. Beyond a doubt
720 These wind blasts, driven from the chill stars of the pole, do
     blow
Against the direction of the river. And the river's flow
Rises from the heat-scorched quarter of the south, way back
Deep in the land of the noonday sun, where nations are burnt
     black.

Or it is also possible that an enormous heap
Of sand obstructs the river's mouth when winds stir up the deep,
And the sea casts sand inside the outlets, silting the egress,
So that the stream's momentum as it flows downhill is less.
It's also possible that at this season more rain falls
730 At the headwaters of the Nile, when northerly Etesian[18] squalls
Corral the clouds up in this area. And you can bet
That once the clouds are driven to the noonday land they get
Thicker and thicker, until the pressure squeezes them together,
Jammed up as they are against tall mountains. Or another
Possibility is that the river's water grows
From deep in Ethiopia's high mountains when white snows,
Melting beneath the all-illuminating sun's warm blows,
Decamp to the fields.

                         And now I shall explain to you what makes
Certain areas 'Avernian',[19] and certain lakes,
740 So pay attention. First of all, they're called by such a word
Because they are anathema to any kind of bird,
Since any bird whose flight-path brings it through their spaces,
     fails
To ply the twin oars of its wings, and furls its feathered sails,
And plummets headlong with its soft and drooping neck to hit
The ground – that is, if solid ground is lying under it –

Or water if it is an Avernian lake stretched out beneath.
There is a place near Cumae like this where the mountains
    breathe
Acrid sulphur fumes and teem with hot springs. And there is
A spot within the walls of Athens, on the Acropolis
Hard by Tritonian[20] Athena's shrine, the Parthenon,   750
Where croaking crows don't ever ply their wings, not even on
Feast days when altars smoke; it is a place they so avoid,
And not because Athena was so bitterly annoyed,
As Grecian poets sing it, when she punished them for spying[21] –
It's from the *noxious nature* of the place that they are flying.
There is another spot like this in Syria, I hear,
A place where even quadrupeds,[22] as soon as they draw near,
Collapse in a heap upon the ground from the place's force, as
    though
They suddenly were sacrificed to spirits down below.[23]

All of these things happen naturally, and it is clear   760
The natural origin whence these phenomena appear –
Lest you suspect these places might be portals into Hell,
Or next, we get it in our heads that here the shades who dwell
In the underworld lure spirits of the living to the shore
Of Acheron; as it is often thought in common lore
That the wing-footed stags, enticing with the breath that rolls
From their nostrils, lure the creeping clan of serpents from their
    holes.[24]
Understand how utterly far-fetched this is, for I
Endeavour in these matters to show you the real reason why.

First of all I tell you, as I've often said before,
The earth possesses every kind of atom in her store: 770
Many are life-giving nutrients, and many more
Can smite us with disease and speed us to our ends. And I've
Shown earlier that different things help different creatures
    thrive,
On account of differing natures, and differing structures and
    sundry forms.
Many a harmful particle enters into the ears, or worms,

Noxious and sharp, into the nostrils themselves, and not a few
There are that we ought not to touch, or which we must eschew
780 With our eyes, or that taste bitter to the tongue.

                                                    Again, it's plain
That there are many elements which trigger pangs of pain
Or discomfort to the human senses, many a loathsome bane.
Firstly, certain trees possess a shade so grave they make
The head of anybody stretched out underneath them ache
As they are sprawled out on the grass beneath the canopy.[25]
There's even, on the mountainsides of Helicon, a tree
Whose blooms have such a noisome stench they'll cause a man's
        demise.[26]
Without a doubt the reason that all of these things arise
From the ground is that the earth has many different seeds
        combined
790 Many different ways, but supplies them separately by kind.
Take the acrid smoke of a snuffed-out night-lamp – when it hits
The nostrils of a person who is prone to falling fits
And foaming at the mouth,[27] it stuns him into sleep. And if
A woman at the time of her menstruation gets a whiff
Of overpowering castor musk, she will let drop the quaint
Needlework from her delicate hands and fall into a faint.[28]
And many other things can make the limbs go limp besides,
And shake the spirit from the inner seats where it abides.
For instance, if you linger too long in a warm bath soon
800 After a heavy meal, how easy to slip into a swoon
Right there in the middle of the steaming tub! Or take
The overwhelming fumes of charcoal, with what ease they
        snake
Into the brain, if we don't drink a preventive glass of water.
And when the body of a man is gripped by raging fever,
The scent of wine will smite him like a sacrificial blow.

Don't we see that sulphur comes right from the earth below
And asphalt with its loathsome stench congeals there? When
        men follow
Veins of gold and silver ore, and delve in deepest hollow

And recess of the earth with iron tools, what fumes exhale
Out of the bowels of Skapte Hyle! And what poisonous bale   810
Breathes from metal mines, what hues and faces, sickly and pale,
It gives to men. Haven't you heard how miners made to slave
By force in such hard labour are sped towards an early grave,
Their store of vital energy sapped out of them? The earth
Therefore exhales all of these vapours, and it breathes them
    forth
Abroad into the open, into the waiting firmament.

It's in this manner also that Avernian places send
Their lethal power up to birds, the fumes that rise on high
Out of the ground to poison that particular patch of sky,   820
So that as soon as any bird comes winging through the air
Into that place, it gets caught in the unseen poison's snare
And drops straight down, right on the very spot from which the
    stream
Of vapour rises, and once it's fallen, the force of that same steam
Robs all its limbs of what is left of life. For it would seem
That first the gasses trigger airy-headedness; then, of course,
Once the bird has plunged right to the poison's very source,
It has to vomit forth its life as well upon the ground,
Since there is such a great supply of evil all around.

It also happens that from time to time this power and breath   830
Of Avernian places knocks away the air between the earth
And the bird, so that it leaves behind an almost empty space,
And when the birds come winging their way directly through
    that place,
Suddenly their flapping falters, and they try to beat
Pinions in vain, the struggle of their wings meets with defeat.
For now when they can get no purchase with their wings at all,
No lift, the downward pull of their own weight compels their
    fall.
Sprawled on the ground in that near-empty spot, they scatter
    their souls
Abroad through all the channels of the flesh, its pores and holes.

840 Consider why well-water's cooler in the summer season:
The earth expands, more porous in the heat, and for that reason
Releases into the air what seeds of heat it might enclose,
And the more the earth breathes out its warmth, the more the
        water grows
Cold lying hidden deep within its bowels. And yet when
All the earth is pinched with cold and it contracts again,
And as it were 'congeals', then by condensing, the earth presses
Into the wells, doubtless, any heat that it possesses.

Hard by the shrine of Ammon, there's a spring that by the light
Of day, they say, is cold, but that runs hot all through the night.
850 Men make too much of the 'miracle' of this spring. Some even
        say
There is a hot sun underground that makes it seethe this way
When night has shrouded earth beneath its terrible pall of
        black.[29]
This explanation is far-fetched, and not on the right track!
For if the sun could not warm up the shallows when it beat
Directly on the surface of the water, with its heat
Blazing from *above* – how can it make the water boil
From *underneath* the earth, below the dense mass of its soil,
And steep it in its warmth? Why, even when the sunshine falls
Directly on a house, it scarcely penetrates the walls
860 With its heat, despite the burning of its rays! So what explains
This spring? No doubt, because the earth around the spring
        remains
Especially porous, and there are many seeds of fire right
In the vicinity of the body of water, so that when the night
Has drowned the land beneath its dewy tide, the earth grows
        chill
To the core, suddenly, contracting. And that is why it will,
As if it were squeezed by a fist, press out what seeds of fire
It has into the spring, and make the temperature rise higher,
And thus the water of the spring's hot to the touch, and steams.
Then once the risen sun has loosened the earth up with its
        beams

Mingled with burning heat, and made the earth porous once
    more,   870
The particles of fire return to where they dwelt before,
And all the water's warmth withdraws into the ground – this
    way
The water of the spring is chilly by the light of day.
Besides, the sun's rays ripple the spring, and shivering waves of
    heat
Rarefy the texture of the liquid as they beat,
So that the waters release whatever seeds of fire they hold,
As we see water that is frozen will give up the cold
Which it possesses, thaw, and loosen up the knots of ice.

There is another chilly spring where oakum, in a trice,
When placed above the waters starts to shoot out flames and
    scorch.   880
And in the selfsame manner, if you set afloat a torch,
It kindles and flickers in the waves, and sails where the breezes
    blow –
No doubt because the spring has many heat seeds, and below,
From the depths of the earth itself, many particles of fire
Must rise through all the water of the spring, and then
    respire
Into the open, released into the air – while yet too few
To make the fountain hot. Besides, a force compels them to
Burst through the water and scatter in the open suddenly,
And convene together just above the surface – as in the sea
There is a spring, the font of Aradus, a spot which wells   890
With sweet water, pushing back surrounds of salty swells,
And many other places in the ocean that supply
Respite in the nick of time to sailors parched and dry
By vomiting fresh water from amidst the waves of salt.
In the same way from this fountain, particles of heat can vault
Into the open air, and scatter abroad. And when they throng
Together onto the length of oakum, or they cling along
The body of the torch, they suddenly ignite with ease,
Since there are also seeds of fire that lurk in both of these.

900 And haven't you noticed – when you bring a just-extinguished
         wick
     Near to a burning night-lamp, how it catches fire so quick
     It lights before it's even touched the flame?[30] And it's the same
     With torches – and many other objects will burst into flame
     Before the fire itself is near at hand and soaks them through.
     So we must understand that's how the fountain functions too.

     But let's move on. I shall begin to tell you by what law
     Of nature lodestone has the power in itself to draw
     Iron to it (that stone the Greeks call 'magnet' from the name
     Magnesia,[31] the city from whose area it came).
910 Men marvel at this stone, because it frequently can make
     A chain of little circles dangle from it. It can take
     As many rings as five or six – you'll see them hanging there,
     Depending in a column, swaying with a breath of air –
     One hanging from another, and each having power in turn
     To hold the next, because the rings through one another learn
     The stone's magnetic properties – the fetters of its force –
     To such extent its powers permeate through things and
         course.[32]
     You must establish many principles before you seize
     A reason for the thing itself in matters such as these,
     And only by a long, circuitous pathway can you find
920 It out – so give the full attention of your ears and mind.

     First of all, from every object visible and showing,
     A stream of particles must be perpetually flowing,
     Particles that strike the eyes and trigger sight. As well,
     There streams from certain objects a continual flow of smell,
     As coolness flows from rivers, heat from the sun, and briny
         spray
     Pours from the ocean waves and gnaws the seaside walls away.
     Various sounds go seeping through the air and never halt.
     And often strolling by the sea, we get a tang of salt
     In the mouth, or when we stand by and we watch someone
         prepare
930 A dose of wormwood, we can taste a bitterness in the air –

So much do various bodies stream from every thing and flow
And scatter in every direction on all sides, and there is no
Delay and there's no respite to disrupt the stream, since we
Sense things sans interruption, and at all times we can see
All things, can smell their odours, and their noises strike the ear.

Now I'll reiterate what I already have made clear
In the first book of my song – *all things are full of emptiness.*
Many different issues hinge upon this knowledge, yes;
But especially for what I'm going to show, it must be fixed   940
Once and for all that everything that we behold is mixed
With void. Firstly, consider how in caves stalactites ooze
With damp, and dribble and drip, and how from every pore we
    lose
Moisture in the form of sweat. The beard grows, and hairs sprout
From each limb over the entire body. And throughout
The whole network of veins, food is dispersed, and this avails
Growth and nourishment down to the toes and fingernails.
We feel that chafing heat passes through bronze, and so does
    cold,
And we can also feel them go through silver or through gold
When we hold a brimming goblet in our hands. And voices leap   950
Through the inner masonry of houses. Odours also seep
Through things, and cold, and the heat of fire that is wont to
    pass
Through the very sturdiness of iron. Even the cuirass
Of the circling heavens can be permeated from without
[By particles of cloud and storm that suddenly come about,]
Or by a pestilence that creeps in from beyond the skies.
Consider also storms that from the atmosphere arise,
Or else to which the vapours of the ground have given birth,
How naturally they are absorbed into the sky or earth,
Since there is nothing that's not made of void woven with
    matter.

It's also true that not all of the particles that scatter
From objects act upon the senses in the same way, nor   960
Is every kind of atom fitted to every substance. For

First of all the sun can bake the ground all hard and dry,
And yet it melts the ice; its rays compel the snow piled high
Upon the mountaintops to thaw, and wax will liquefy
Placed in his heat. Fire also makes bronze liquid, and it melts
Gold, but at the same time shrivels meat and shrinks up pelts.
Take water, how it tempers iron hot from the forge, but then
Makes flesh and skins hardened by heating pliable again.
970 The wild olive is such a treat to bearded nanny goats
   You'd think that they were guzzling ambrosia down their
      throats,
   And yet to man there is no bitterer foliage on earth.
   A pig loathes any ointment, and gives marjoram wide berth,
   For what seems soothing and refreshing to ourselves, instead
   Is, to bristly pigs, a baleful poison that they dread.
   But muck, on the other hand – to us the most disgusting stuff –
   Is such a joy to pigs that they can never get enough
   Of wallowing and rolling in it.

                              But before I press
980 Ahead with the question at hand, there is one thing I must
      address:
   Since there are many pores in things, these channels must
      possess
   A range of differences, and thus each kind of thing displays
   A certain character due to its different passageways.
   Indeed, all creatures are equipped with various senses, each
   Of which perceives its own particular stimulus. Sounds reach
   Us by the conduit of one sense, and flavour by another,
   While yet another sense is permeated by an odour.

990 Besides, we see that one thing slips through stone, another seeps
   Through wood, another goes through gold, and still another
      creeps
   Through silver or glass: for instance, images can stream through
      glass,
   And warmth trickle through silver. And we perceive one thing
      can pass

More quickly than another through the same substance. No
    doubt
The nature of the channels is what makes this come about,
Since they are different in myriad ways, as I have shown before,
Due to their sundry natures, whether they have less or more
Void in their weave. And when all of these principles, therefore,
Have been proved and established for us, the one thing to
    remain –
The question now before us – becomes easy to explain:
Namely, to make clear, extrapolating fact from fact,   1000
Why lodestone acts on iron with the power to attract.

Firstly, there must be a quantity of seeds that flows
Out from the stone, a streaming wave of seeds that, with its
    blows,
Knocks away the particles of air that intervene
Betwixt the iron and stone, and when the span that lies between
Is left empty, and a vacuum opens in that place,
The particles of iron slip forward to the empty space,
Tumbling in, all joined together, so the iron ring
Follows en masse, moving as one. Nor does anything
Consist of elements more closely tangled up and tight,   1010
Bound more coherently than iron's cold and bristly might.
And since they are so tightly tangled, it is no surprise
That when, out of the iron, quantities of seeds arise,
They cannot slide into a vacant space without the ring
Coming along with them, which the ring does, following
Until it reaches the lodestone, and it forges a connection
By means of unseen clasps. It can move in any direction
Because, wherever a vacuum opens up, either to the side
Or up above, the neighbouring particles at once will glide
Into the void, herded by blows that pelt from everywhere.   1020
They are not able, *on their own*, to rise into the air.

Here's something that facilitates the movement even more
And helps the ring along: as soon as the air that lies before
The ring becomes more thin and rarefied – and thus the space
That lies before the ring becomes a voider, emptier place –

All at once, the air that lies behind the ring will shove
It forward from the back, and thus compel the ring to move.
The surrounding air forever rains its blows on everything,
But the reason that, at this particular time, it drives the ring,
1030 Is the ring can be received on one side into empty space.
This same air nimbly infiltrates the many passageways
Of iron down to its tiniest parts and pushes it, as gales
Of wind impel a ship along by bellying its sails.
Lastly, everything contains some air within its bounds,
Because all things are porous, and because the air surrounds
And touches every object. So air lurking deep within
The iron is tossing and turning in a constant, ceaseless motion,
Without a doubt the air is beating against the ring and drives
1040 It from within, and the ring itself is borne along and dives
Towards the vacancy in the direction of its momentum.

It also happens sometimes that the iron recoils from
The lodestone, alternately attracted to it and repelled.
Why, I've seen Samothracian irons[33] jump up, and beheld
Iron shavings in a brazen basin rage and seethe
When this magnetic stone is situated underneath,
So jittery and jumpy are they to escape the stone.
But the blame for this disharmony lies with the bronze alone
Placed between them. Doubtless because the stream of seeds that
        spills
1050 From the bronze takes hold of the open channels of the iron and
        fills
Them up before the current from the stone does, which no more
Can find the pathways that it used to navigate before,
So the stone must beat against the iron's fabric with its tide,
Repelling and agitating filings on the other side
Of the bronze that it would otherwise attract.

                                                Thus it should prove
No surprise that lodestone does not have the power to move
Every kind of object. For there are some things that stand
Fast due to their weight, for instance gold. On the other hand,

Other objects, being porous, allow the wave to pass
Right through them without interruption (wood is in this class),   1060
And so the current is unable to propel their mass.
Iron's nature lies somewhere between the two, and so,
Once it has taken on small particles of bronze, can go,
Driven onward by the stones and their magnetic flow.

But this phenomenon is not unique or even strange;
Indeed, I have at my disposal an entire range
Of examples where two stuffs are fitted solely for each other,
As you will notice only lime can mortar rocks together,
And wood planks joined with ox-glue gape at the grain before
    the hold
Of the glued joints will loosen. Then, the juice of the vine is bold   1070
To mix itself with spring-water, when neither sinking pitch
Nor floating olive oil is able to. Also, the rich
Wine-dark stain of mollusks[34] will so totally unite
With wool, you can't divide them out again, try as you might –
Not even with the flood of Neptune could you wash it free,
Not though you rinsed the fleece with all the waters in the sea!
And is not chrysocolla[35] the only thing that solders gold
To gold, and isn't tin the only solder that will hold
Copper to copper? I could go on and on – there are many other
Examples of this kind we can discover. But why bother?   1080
Why should you go the long way down a path so roundabout,
And why should I belabour the argument and drag it out?
It's better to cover much in a short space. And so, in fine,
When it happens that the textures of two substances align
So that the solid places of one can meet up with the hollows
Of the other, and vice versa, then a perfect union follows.
There also might be substances that are linked up together
By a system of little hooks and eyes that clasp one to the other.
With iron and lodestone, evidently, this is what applies.

Now I'll explain the causes of disease: whence does it rise,   1090
So of a sudden a pestilence can kindle the demise
Of humankind and of the legions of beasts. For first, as I've
Taught earlier, there are many seeds of things we need to thrive;

But on the other hand, there are many seeds that swarm around
Engendering sickness and death, and when by chance these seeds
    abound,
They throw the heavens into turmoil and infect the air.
And all this onslaught of pestilence comes either from
    somewhere
Beyond the world, as mist or clouds, through the heavens
    overhead,
1100 Or it gathers and arises from the earth itself instead,
When the sodden ground is rotten with mud, lashed by the sun's
    rays
And unseasonable rains. Don't we see how someone who strays
Far from hearth and homeland is afflicted by the strange
Climate and water, because they undergo so great a change?
Just think how different the climate where the Britons dwell
Must be from Egypt's, where the axis pivots – think, as well,
How climes of Black Sea and Cadiz must differ – or way back
Down to where the sunshine bakes the skin of nations black.
1110 And as we see four different climates in the four directions
Of the four cardinal winds, and as men's features and
    complexions
Differ widely, ailments also vary by nation. While
Elephantiasis strikes only near the river Nile
In the midst of Egypt, in Attica, instead, problems arise
With the feet, and in Achaea, what's afflicted are the eyes.
Thus different places are hostile to different members of the
    frame;
The variety of airs in different places is to blame.

So when a sky bestirs itself, one that is alien
1120 To us – and an unwholesome air starts slowly creeping in –
Slinking in the manner of clouds and mists – wherever it crawls,
It wreaks havoc and alters everything. If it befalls
That such a miasma enters our own climate, it will change
Our sky with its contagion, making it like itself and strange
To us. So of a sudden a new plague or pestilence drops
Down upon the waters, or even settles on the crops

Or other source of food for men or fodder for beasts, or floats
Suspended in the air itself, so that we breathe its motes
Together with the air that we inhale, and take it deep
Inside our flesh. And in the selfsame way, plagues often sweep   1130
Through herds of cattle; now distemper strikes the dawdling
   sheep.
Nor does it make a difference whether it is we who range
To another climate, hostile to ourselves, and we exchange
Our mantle of sky, or whether it is, rather, Nature who
Imports a sky contaminated, a climate that is new
To us who, unaccustomed, are more vulnerable to attack.

This is what caused the fatal tide of pestilence years back
In Athens,[36] land of Cecrops, that blighted fields, and left no trace
Of traffic on the roads, and made a ghost-town of the place.   1140
From deep in the heart of Egypt it uprose, and crossed the vast
Expanse of air and swimming plains, until it fell, at last,
Wholesale upon the people of Pandion. Wave after wave,
In legions, they were handed over to sickness and the grave.

The symptom first to strike was fiery fever in the head,
And both eyes, burning hectic bright, were all shot through with
   red.
The throat as well would sweat with blood, all black within.
   And stung
With sores, the pathway of the voice would clog and choke. The
   tongue,
Interpreter of the mind, oozed pus, and, made limp with the
   smart,
Was too heavy to move, and rough. Thence the disease would
   start,   1150
Passing the gullet, to fill the chest, and flood the heavy heart
Of the afflicted, and then, indeed, all of the gates of Life
Began to give. From the open mouth, there would exhale a rife
Stink, like the stench of rank unburied corpses left to rot.
And then all of the powers of the mind and body, brought
To the very brink of doom, began to flicker. Mental strain
Ever danced attendance on intolerable pain;

Pleas mingled with moans. Ceaseless retching, lasting day
1160 And night, was ever causing seizure and cramp, and wasting
        away
The strength of men already racked with suffering and worn out.

And yet if you laid hands upon a sufferer, you did not
Feel that the surface of his body was excessively hot;
Rather, to the touch it felt slightly warm – though you
        discerned
The skin was red and broken out with blisters as though
        burned,
Or else as when the Sacred Fire covers all the frame.
Yet on the inside, men were burning to the bone. A flame
Blazed in the stomach as if in an oven. There was no stuff
1170 That you could lay upon their skin, flimsy or light enough –
Only coolness and air were ever bearable. Some gave
Their feverish limbs to cooling streams, and plunged into the
        wave
With naked bodies. Many tumbled down into a well
Headlong from a height, mouths gaping open as they fell.
A parching thirst that drenched them through and through and
        would not stop
Made gushing floods of water seem no better than a drop.

Nor was there any respite from the pain. The bodies lay
Worn out. Even Medicine did not know what to say,
1180 But mumbled in terror, haunted by those eyes that ever keep
Rolling, ablaze with fever, and are never closed in sleep.

At that point there were many other signs that death was near –
A mind that was deranged with melancholy, mad with fear,
A frowning brow, an angry and ferocious countenance,
The ears a-throb with tintinnabulation; rapid pants
Of shallow breath, or deep and ragged gasps. A stream of bright
Sweat ran trickling down the neck. The phlegm was scant and
        slight,
Salty, and tinged with a yellow hue, and even the hacking cough
Could scarcely bring it up the throat. The hands never left off

Their twitching, and the limbs never stopped shaking. By and
     by,   1190
A chill stole creeping from the toes, and as the end drew nigh,
The nostrils became pinched, the nose as well grew sharp and
     thin,
The eyes sunk in their sockets, temples hollowed, and the skin
Went cold and hard, the mouth was grimaced in a rigid grin,
The forehead taut and stretched. And not long afterwards, limbs
     lay
Stiffened in death. Usually at the gleam of the eighth day,
Or by the shining of the ninth, they would draw their last
     breath.

The odd survivor – for there were some – had not cheated death:
For him, a wasting away and slow demise still lay in wait.   1200
Either, running sores and black flux from the bowels, or spate
Of corrupted blood poured through the nose, along with a
     throbbing head –
The patient's might and main ebbing away with what he bled.
And if the haemorrhaging of foul blood did not leave him dead,
The plague proceeded to the limbs and muscles, and progressed
Even to the genitals. Some people were possessed
With such grave terror at the door of Death that with a knife
They managed to castrate themselves, and so hang on to life.
Many lingered in this world sans hands or feet. Some lost   1210
The light of their eyes. This was the price such dread of dying
     cost.
Some fell into a deep forgetfulness, and lost all store
Of memories, and did not know their own selves any more.

Though corpses piled on corpses lay in heaps upon the earth
Unburied, yet the tribes of birds and beasts gave them wide
     berth
And sprang away in order to escape the noisome stink.
Or else they, having tasted of the flesh, at once would sink
Down dead – though there were hardly any birds to speak of,
     nor
Did the dismal tribes of wild beasts venture from the forest, for   1220

The animals fell sick in droves and perished. But the clan
Of dogs was hardest hit,[37] the true and steadfast friend of Man:
Strewn on every corner, life wrenched from them by the might
Of the plague, they still did not give up the ghost without a
     fight!

One unattended funeral raced another to the grave.
There was no common remedy that would be sure to save –
For what had given one the breath of life, so he could sigh,
Thankful to behold the shining regions of the sky,
Proved fatal to another and dispatched him to his doom.

1230 But in these matters, what was saddest and most cause for gloom
Was that, when someone saw the plague upon him, he would
     start
Thinking like a man under sentence of death, and would lose
     heart,
And lay there listlessly, his mind sunk deep in morbid thought,
And dwelling on his death, gave up his spirit on the spot.

At no time did the greedy disease let up. It caught and spread
From one man to another, as though they were so many head
Of fleecy sheep or cattle. This is what piled deaths up thick:
That if a man shirked visiting his own kin fallen sick,
1240 For his excessive lust for life and dread of death, he'd learn
The fatal price of Negligence, neglected in his turn,
So shortly afterwards he came to meet a shameful end,
No one by his side to aid him; while those who *did* attend
The ill, on the other hand (their sense of duty left no choice),
Compelled by pleas for pity mingled with Reproach's voice,
Contracted the disease run down by labour without rest
In the sickroom. And that's how death took the noblest and the
     best.

[...]
People struggled to bury legions of their kinfolk, heaping
One funeral on another, and returned wrung-out with weeping,

Whereupon many took to bed from sheer grief. There was such
Widespread woe that there was none who had not felt the touch   1250
Of pestilence or death or mourning at this point.

                                        By now,
The shepherd, herdsman and the sturdy helmsman of the
    plough,
All were fading away. Their bodies lay crammed in the back
Of cottages, betrayed to death by pestilence and lack.
Sometimes you'd see the lifeless corpses of the parents piled
Upon their lifeless children, or vice versa, you'd see child
After child upon their perished parents lay their small lives
    down.

Mostly this disease flowed from the country to the town.
For the stricken throng of country folk that poured in
    helter-skelter,   1260
From every quarter, carried it with them, packing every shelter
And public space until the death-toll mounted all the more
By cramped conditions and stifling heat that slew them by the
    score.

Many writhed on the roads, prostrate with thirst, or they would
    sink
Beside the fountains, choked from life by their great greed to
    drink.
And many bodies you would notice scattered in plain view,
Limbs dangling from half-dead trunks, in square and avenue,
Caked in squalor, clothed in rags and only skin and bone,
About to perish from their body's filthiness alone,   1270
Already as good as buried in their putrid sores and dirt.

Then Death had filled all shrines with congregations of inert
Corpses; all the temples of the Holy Ones were weighed
Down with cadavers, places where the visitors had prayed
In pressing throngs packed in by sacristans. For at this hour
The worship and the awe of gods had lost most of their power –

The present grief was overwhelming. No one any more
Observed the rites of burial they had observed before,
1280 For the whole populace was thrown in disarray and cowed.
Each mourner buried his dead just as the time and means
    allowed.
Squalid Poverty and Sudden Disaster would conspire
To drive men on to desperate deeds – so they'd place on a pyre
Constructed by another their own loved-ones, and set fire
To it with wails and lamentation. And often they would shed
Much blood in the struggle rather than desert their dead.[38]

Notes

BOOK I

MATTER AND VOID

1. *Venus*: Goddess of love, and, for Lucretius, the symbol of creativity, sexual energy and nature.
2. *Pleasure*: Latin *voluptas*, Greek *hedone*, a key tenet of Epicurean philosophy; its presence at the opening of the poem (in the original, the last word of the first line) is emphatic.
3. *airy tribe of birds . . . through the heart*: Cf. Chaucer's prologue to *The Canterbury Tales* (lines 9–11):

> And smale foweles maken melodye,
> That slepen al the nyght with open yë
> So priketh hem nature in hir corages –
> [so does Nature pierce them in their hearts]

Also see Edmund Spenser's incomparable rendition of the proem in *The Faerie Queene*, 4.10.44:

> Great *Venus*, Queene of beautie and of grace,
> The ioy of Gods and men, that vnder skie
> Doest fayrest shine, and most adorne thy place,
> That with thy smyling looke doest pacifie
> The raging seas, and makst the stormes to flie;
> Thee goddesse, thee the winds, the clouds doe feare,
> And when thou spredst thy mantle forth on hie,
> The waters play and pleasant lands appeare,
> And heauens laugh, & al the world shews ioyous cheare.

4.   *Mars*: Lucretius uses the archaic form Mavors. The pairing of Mars and Venus suggests the Empedoclean notion of Love and Strife as the opposing forces of the universe; see Introduction, p. xii.

5.   *so often lies . . . your lips*: Cf. Byron's *Childe Harold's Pilgrimage* (4.51):

> In all thy perfect Goddess-ship, when lies
> Before thee thy own vanquish'd Lord of War?
> And gazing in thy face as toward a star,
> Laid on thy lap, his eyes to thee upturn,
> Feeding on thy sweet cheek! while thy lips are
> With lava kisses melting while they burn,
> Shower'd on his eyelids, brow, and mouth, as
>   from an urn?

6.   *hour of need*: Rome during Lucretius' lifetime was in a constant state of unrest, the Republic in its death throes; *hoc patriai tempore iniquo* – at this critical moment for the country – might also refer specifically to the period from the second Catilinarian conspiracy in 63 BC to the First Triumvirate in 60 BC, or to Memmius' governorship of 58 BC.

7.   *For godhead . . . by anger*: Some feel these five lines are misplaced – they also occur in II.645–51 (in the Oxford Classical Text II.646–51).

8.   *the first . . . was Greek*: Epicurus, although Leucippus and Democritus had laid the foundations for his atomist vision of the world.

9.   *Religion*: I am using Religion and Superstition interchangeably in these passages. The Latin is *religio*. Lucretius is not against piety but the evils of organized religion.

10.   *The Virgin's altar . . . Iphigenia*: The virgin (here, the Virgin of the Crossroads) is Artemis. A tripartite goddess, she is the virgin goddess of the hunt, the goddess of childbirth, and is associated with Hecate, goddess of the underworld and black magic. (The Virgin of the Crossroads represents this last guise.) Offended by the slaughter of a sacred hind, Artemis caused contrary winds to prevent the Greek fleet from leaving the harbour at Aulis and setting sail for Troy. To appease the goddess, the Greeks sacrificed Iphigenia, daughter of Agamemnon, leader of the Greek expedition. She was told she was to be married to Achilles and was led as a bride to the altar.

11.   *fillet*: Latin *infula*, a red and white band of tufts of wool that marked the sacrificial victim.

12. *So potent was Religion in persuading to do wrong*: Line 101 is one of the most famous lines in the poem: *tantum religio potuit suadere malorum*. Voltaire believed it would last as long as the world. (See also Introduction, p. xi.)

13. *waters yield . . . change places with each other*: The Platonic theory of spatial displacement (see the *Timaeus*, Plato 79b) which was accepted by Aristotle and the Stoics, was that movement was possible without void because all the moving parts (as of water) could simultaneously shift position.

14. *the two*: Void and substance.

15. *rape of Helen . . . war*: Reference to the abduction by the Trojan prince Alexander (aka Paris) of Helen, wife of king Menelaus of Sparta, which sets off the Trojan War.

16. *vintage . . . water's poured in*: In the ancient world, wine was usually served mixed with water.

17. *air and water, earth and fire*: The four 'roots' or elements of Empedocles, from which he thought the universe composed, combined and separated under the respective sways of the forces of Love and Strife. Each root in turn was taken by different philosophers to be the basis for existence. See Book V, note 16.

18. *each species . . . stays the same*: This seems to approach the notion of genes.

19. *Heraclitus*: See Glossary. He was known as 'the dark one' because of the obscurity of his writings.

20. *they*: Philosophers of the doctrine that all comes from fire.

21. *as that man claims*: Heraclitus.

22. *homoeomery*: Of like parts; things are infinitely divisible into parts like each other and like the whole (keep dividing a hair and you will have smaller and smaller pieces of hair). All substances contain a portion of all other substances, though whatever substance predominates determines what a thing is. Flesh can be made by eating bread because bread contains tiny portions of flesh in it, and so on.

23. *[. . .]*: For this lacuna Bailey in his Commentary gives the sense 'must consist of things unlike themselves, which in their turn must contain things unlike themselves'.

24. *maple . . . flame*: The anagrammatic pun in Latin is on *lignis* (wood) and *ignis* (fire).

25. *elements . . . salty tears*: Lucretius' favourite rhetorical device, *reductio ad absurdum*. If everything is made of miniature copies of itself, then our atoms must surely be capable of emotion!

26. *everything . . . certain people claim*: Principally the Stoics (the main

rivals of the Epicureans), although the notion that all tends to a centre was also advanced by the Old Academy and the Peripatetic philosophers, the followers of Plato and Aristotle respectively.

27. *How can this be if everything of earth tends to the middle?*: The lacuna begins here. I have supplied this line to finish the sense of the two previous lines. Lines 1094–1101 were lost when a page of the Archetype manuscript (from which all copies descend) was damaged. The other side of the same page also suffered damage – the ends of lines 1068–75, where Munro's restorations have been followed.

BOOK II

THE DANCE OF ATOMS

1. *palace*: See the description of the palace of the Phaeacians in Homer, *Odyssey* 7.100–102:

> Here, too, were boys of gold on pedestals
> holding aloft bright torches of pitch pine
> to light the great rooms, the night-time feasting.
>> (trans. Robert Fitzgerald, 1961)

2. *Field of Mars*: The Campus Martius was part of the flood plain outside the walls of Rome lying in the bend of the Tiber. It was the site of military exercises.

3. *fleet of ships spread far and wide*: In mock naval battles.

4. *Passing on life's torch, like relay runners in a race*: The metaphor comes from the Athenian torch relay-race (Lampadedromia), performed on horseback, in honour of the Thracian goddess Bendis. Plato uses the race as a metaphor for the handing on of life from one generation to the next in his *Laws* (776b). This passage in Lucretius gives the title to Sir Henry Newbolt's popular 1897 poem, 'Vitaï Lampada', which closes:

> This they all with a joyful mind
> Bear through life like a torch in flame,
> And falling fling to the host behind –
> 'Play up! play up! and play the game!'

5. *[. . .]*: The lacuna here is probably lengthy – maybe as many as 52 lines, if a page of the Archetype was missing.

6. *certain people*: Lucretius probably has the Stoics in mind here, for whom Nature is governed by divine Reason (God).

7. *swerve . . . change of course*: We do not have any extant writings of Epicurus that set out his theory of the atomic swerve; our fullest account is here in Lucretius. It was on the basis of this slight swerve that Epicurus explained freewill. The Epicurean emphasis on free-will is very much at odds with the Stoic resignation to Fate.

8. *saffron's . . . on the stage*: Several Romans mention the sprinkling of the stage with saffron water. See Horace (Epistle 2, I.79, 80): 'If I were to question whether a play of Atta's keeps its legs or not admidst the saffron and flowers . . .' (Horace: *Satires, Epistles and Ars Poetica*, trans. H. Rushton Fairclough, 1926), and Ovid, *Ars Amatoria* 1.103, 104: 'Then no curtains hung over a marble theatre, nor were the stages ruddy with liquid saffron' (trans. AES).

9. *dried wine lees or scabwort*: These are cream of tartar, with an acidic flavour, and scabwort (elecampane), a bitter herb. Romans used both as condiments.

10. *Now that I've demonstrated this*: A passage may be missing, or there is a gap in the argument; 'this' would seem to refer to a proof that the size of the atoms is limited.

11. *murex . . . Thessalian tide*: Meliboea, a town on the coast of Thessaly, produced an expensive purple dye from the shellfish murex.

12. *song the swan sings*: Legendary for its beauty in ancient times (though in actuality swans do not sing); swans were supposed to sing especially beautifully at the point of death (hence our idiom 'swan song').

13. *snaky-handed elephants . . . so rare*: The Romans first encountered war elephants when Pyrrhus (king of Epirus, northwest Greece) invaded Italy (280 BC). The Roman cavalry was routed by the strange and frightening spectacle. Elephants were first displayed in Rome as part of the triumph when Pyrrhus was defeated in 275 BC. Romans later encountered elephants when Hannibal famously crossed the Alps with a cohort in the Second Punic War (218 BC; see Book III, note 18). In 55 BC Pompey staged a games (see Book VI, note 8) in which twenty elephants were slaughtered by spearmen. The sight of the animals, who showed an almost human understanding of their plight, being so butchered, excited the pity and distaste of the crowd. Cicero, who was present, wrote to a friend: 'What amusement can it be to a cultivated person when either a feeble man is torn to pieces by a mighty brute or a noble beast [elephant] is run through with hunting spears?' (*Ad Familiares* VII, 1). See also Book V, note 33.

14. *mighty sea . . . oars*: For the asyndeton (lack of conjunctions), cf.
      'the sea, sweeping over the rolling wreck, made a clean breach, and
      carried men, spars, casks, planks, bulwarks, heaps of such toys, into
      the boiling surge', Charles Dickens, *David Copperfield* (ch. 55,
      'Tempest').

15. *Great Mother*: Cybele (see Glossary), known as the Idaean Mother
      (611) because Ida, a mountain range in Mysia, Asia Minor (modern
      Turkey), was the ancient seat of her worship. Another description
      of Cybele and her cult can be found in the work of Lucretius'
      contemporary Catullus (Catullus 63).

16. *songs of ancient Greek bards*: Greek poets who describe Cybele
      with her lion chariot include Pindar, Sophocles and Euripides.

17. *eunuchs*: Priests of Cybele were eunuchs (615; Oxford Classical
      Text 614) because they castrated themselves, in honour of Attis, a
      youth beloved by Cybele, who went mad and castrated himself
      with a sharp rock. The blades in Cybele's procession represent the
      castration.

18. *Dictaean Curetes . . . baby Jupiter's wails*: Dicte was the mountain
      in Crete where baby Zeus (Jupiter) was reared. The god Saturn
      knew he was destined to be supplanted by one of his children. Thus,
      whenever his wife Rhea had a baby, he swallowed it. But when Zeus
      was born, she gave her husband a swaddled stone to swallow instead
      and spirited the baby to Crete. The Curetes were attendants of Rhea,
      who drowned out his cries with the clashing of their weapons. (They
      are often confused, as here, with the Corybantes, eunuch priests of
      Cybele, who followed her with wild dances and music. The fact that
      both Rhea and Cybele are closely associated with a Mount Ida,
      Rhea with the one in Crete and Cybele with the one in Asia Minor,
      added to the confusion. Both goddesses are known as 'mother of
      the gods' and associated with fertility and orgiastic rituals.)

19. *gleam of white . . . any colour you could name*: Ovid in his *Metamor-
      phoses* (2.534 ff.) tells the story of the raven who was once a
      snow-white bird but was turned black for being a tattle-tale. The
      scarcity (i.e., non-existence) of black swans was proverbial in the
      ancient world, as Juvenal's famous line suggests: 'Rara avis in terris
      nigroque simillima cygne' ('A rare bird upon the earth, and very
      like a black swan', Satire 6.165). The black swans of Australia were
      of course unknown in Europe at this time.

20. *spikenard bloom*: An exotic Eastern plant, now identified as the
      Nardostachys jatamansi of Northern India, from which the ancients
      prepared a costly and aromatic ointment or oil.

21. *those who claim*: Probably referring to Anaxagoras and his theory of homoeomery (see Glossary; I.830 ff. and note).

22. *atoms ... drawn to what's the same*: The idea of like attracting like seems to be derived from Empedocles' concept of affinities rather than Epicurus.

23. *once gave birth ... monstrous beasts*: A concept perhaps suggested by actual fossil bones.

24. *lowered on a rope of gold*: The image of the golden rope from heaven comes from Homer, *Iliad* 8.19–22, where Zeus says: 'Hang a gold cord down from heaven, and all you gods and goddesses take hold of it: but you could not pull Zeus, the counsellor most high, down from heaven to the ground, however long and hard you laboured' (trans. Martin Hammond, Penguin Classics, 1987). The image was later taken up by Plato. The idea of a cord or chain lowered from heaven may also suggest the Stoic notion of the unbreakable chain of causality.

25. *nor that mortals came ... smash the rocks*: Here Lucretius is probably challenging the notions of the philosopher Anaximander (*c.* 610–545 BC), who held that life arose spontaneously from mud warmed by the sun, and that men were first gestated until adolescence inside fish-like creatures, since they were too vulnerable to fend for themselves.

BOOK III

MORTALITY AND THE SOUL

1. *You*: Epicurus.

2. *The gods appear ... laughter of its light*: Lines 17–22 imitate Homer's *Odyssey* 6.42–6, and are in turn imitated by Tennyson in his poem *Lucretius* (104–10):

> The Gods, who haunt
> The lucid interspace of world and world,
> Where never creeps a cloud, or moves a wind,
> Nor ever falls the least white star of snow,
> Nor ever lowest roll of thunder moans,
> Nor sound of human sorrow mounts to mar
> Their sacred everlasting calm!

3. *when they claim ... air*: This is aimed at those who arbitrarily accepted one of the materialist explanations of the soul, not aimed

at a specific philosophical school, although Empedocles held the seat of thought was blood around the heart, and Anaximenes and Diogenes of Apollonia thought the soul was air.

4. *black livestock*: Romans offered sacrifices to the Manes, the spirits of their dead ancestors. As a rule, celestial gods received fair or white victims, and spirits and deities of the underworld received black victims. Cf. Virgil, *Aeneid* 6.243–54, where Aeneas sacrifices four black bullocks, a black-fleeced lamb and a barren cow to various infernal powers before entering the underworld.

5. *Avarice . . . Ambition*: See Book V, note 5.

6. *Death's gate*: Lucretius' metaphor here is made literal and expanded by Virgil in his description of the entrance to Hell in the *Aeneid* 6.273–6:

> Right at the entrance, at the mouth of Hell,
> Grief and vengeful Cares have made their beds,
> There pale Diseases and sad Old Age dwell,
> And Dread, and Hunger who gives bad advice,
> And Ugly Want . . . (trans. AES)

7. *harmony*: Used here in its (ancient) Greek sense – not of chords or notes played together, but of the tuning of an instrument. Literally, a fitting together, an agreement or concord. In Plato's *Phaedo* Simmias tells Socrates: 'our body is kept in tension, as it were, and held together by hot and cold, dry and wet, and the like, and our soul is a blending and attunement [harmony] of these things, when they're blended with each other in due proportion' (trans. David Gallop, 1975). The theory was probably Pythagorean but was taken up by later philosophers.

8. *spirit*: In this section 'spirit' translates *anima* – not really 'soul', but the 'vital principle', as opposed to *animus*, the 'intellect'. The soul is made up of a combination of the two.

9. *the entire body . . . give way*: These symptoms of fear are strikingly similar to the physical symptoms of love in Sappho 31: 'my tongue has snapped, at once a subtle fire has stolen beneath my flesh, I see nothing with my eyes, my ears hum, sweat pours from me, a trembling seizes me all over, I am greener than grass' (*Greek Lyric I: Sappho and Alcaeus*, trans. David A. Campbell, 1982).

10. *mound of stones . . . Won't budge*: Stones won't move because of their weight, ears of wheat because they are 'rough' and catch on one another.

11. *soul*: Mind and spirit together.

12. *threefold*: I.e., breath, heat and air – *aura* (breath or wind) and *aer* (air) here not the same thing.

13. *Like many other qualities*: Such as heat, colour, movement.

14. *opened doors*: The idea that eyes were literally the windows (or doors) of the soul was adopted by the Stoics.

15. *when wine . . . drunken spree*: The predictable stages of drunkenness were a common subject for ancient writers. See Euboulus, fragment 94, in the voice of Dionysus, god of wine:

> For men of moderation, I mix only three kraters of wine:
> The first they quaff for health, the second for love and pleasure,
> The third for sleep;
> Having downed that, sensible men shuffle home.
> I wash my hands of the fourth: it belongs to excess.
> The fifth is for raised voices. The sixth, carousing,
> The seventh for black eyes. The eighth calls the constable.
> The ninth is for depression. The tenth belongs to madness
> and mayhem. (trans. AES)

16. *biting humour . . . spirit back again*: Epilepsy was known as the 'Sacred Disease' in the ancient world, but a treatise ascribed to Hippocrates (*c.* 430 BC), in an important step towards the differentiation of science from religion, debunked the notion that this disease was sent by the gods, and instead found its cause in an excess of phlegm, one of the (ancient Greek) humours. Other writers disputed which humour was to blame; Lucretius does not specify.

17. *Hyrcanian hounds*: A particularly territorial and fierce breed from the southeast shores of the Caspian. See also note 19.

18. *we did not . . . feel anxious . . . shock of war*: Reference to the Second Punic War (218–202 BC), fought between Rome and her rival, the North African city-state of Carthage. This war was considered by the Romans one of the most significant crises in their history. Hannibal famously invaded Italy, having crossed the Alps with his army and elephants. In fact, the Carthaginians attacked from almost every quarter: north, south, west, and through their allies, from the east. Carthage was defeated, but Rome remained traumatized, and went on to destroy Carthage in the Third Punic War.

19. *corpse . . . savage beasts*: To have the corpse pulled apart by animals may refer to the custom of the Magi (as well as, of course, to the ignoble fate of corpses on the battlefield). Cicero mentions that the Magi, the priestly caste of the Zoroastrians, only buried their dead

after they had been torn apart by beasts, and that in Hyrcania, hounds were kept at public expense for that purpose. He also notes that 'the Egyptians embalm their dead, and keep them at home; the Persians even coat their dead in wax before burying them, to preserve the bodies as long as possible' (*Tusculan Disputations* I.108, trans. AES).

20. *No more happy welcome-home . . . race to kiss you*: This passage reflects the standard laments of mourners. It was famously imitated by Thomas Gray in his *Elegy Written in a Country Churchyard* (1751):

> For them no more the blazing hearth shall burn,
>   Or busy housewife ply her evening care;
> No children run to lisp their sire's return,
>   Or climb his knees the envied kiss to share.

21. *Tantalus*: There are two accounts of the punishment of Tantalus in the underworld. The more famous, from Homer and other authors, is of his being 'tantalized' by fruit he cannot reach and water that recedes when he tries to drink. The second, from Pindar's First Olympian Ode, involves his being perpetually menaced by a rock hanging over his head. Lucretius chooses the latter version.

22. *maidens in their bloom*: The Danaïds, fifty daughters of Danaus, forty-nine of whom murdered their husbands. The murderous brides were punished eternally in the underworld by having to fill leaky pots with water.

23. *Traitor's Rock*: The Tarpeian rock on the Capitoline Hill from which traitors were cast.

24. *Even Ancus . . . upon the light*: A line essentially quoted from Ennius, *Annales* III, 137 (149). Ancus was the (legendary) fourth king of Rome.

25. *he*: Xerxes, king of Persia, who built a pontoon bridge over the Hellespont in 480 BC, on his way to invade Greece.

26. *Scipio*: Probably Publius Cornelius Scipio, the elder Africanus, who defeated Hannibal at Zama (North Africa) in 202 BC.

27. *Epicurus*: This is the only place he is mentioned by name by Lucretius.

BOOK IV
THE SENSES

1. *I've taught you ... from which it came*: A passage that may show the unrevised nature of the poem, which would probably have been cut in revision. It is perhaps the original preface, later superseded by the preceding lines.

2. *awnings ... fluttering glow*: In Lucretius' day, Roman theatres were temporary wooden structures. (Pompey's stone theatre was built in 55 BC.) Awnings to shade spectators (performances were in the heat of the day) were first introduced in 78 BC. Sockets for the masts which supported the awnings are still visible in the stone theatres.

3. *glass*: I am occasionally using 'glass' and 'looking glass' for 'mirror'. Mirrors in antiquity would have been metal, not glass with silvered backing (a relatively recent invention).

4. *overpowering lad's-love*: Southernwood (or old man), a shrubby plant from the southern Mediterranean, with a fragrant smell but sour taste; cf. Edward Thomas's poem 'Old Man' (1914):

    > Where first I met the bitter scent is lost.
    > I, too, often shrivel the grey shreds,
    > Sniff them and think and sniff again and try
    > Once more to think what it is I am remembering,
    > Always in vain.

5. *as if in one continuous team*: As in a team of oxen.

6. *switched places with each other*: The 'problem' of why mirrors reverse left and right (but not up and down) continues to be of interest; see Jim Holt, 'The looking-glass war: why don't we see ourselves upside down in the mirror?', *Lingua Franca* 10 (8) (November 2000).

7. *Or take a mirror ... curved shape teaches*: Lucretius seems to have in mind a horizontal concave mirror, rather than a bowl-like mirror. But even looking into a soup spoon provides a similar experiment – the image appears upside-down, but right and left are not reversed.

8. *images return ... angle of attack*: Lucretius seems to be expressing that the angle of reflection equals the angle of incidence.

9. *like a thread ... spun into a fire*: Possibly alluding to a proverb about futility. Cf. Plato's *Laws*, 780c: 'that's what makes the legislator "card his wool into the fire", as the saying is, and make so many

efforts fruitlessly' (trans. by Trevor J. Sanders, Penguin Classics, 1970).

10. *no deeper than a finger*: Lucy Hutchinson, who englished the *De Rerum Natura* in the 1640s, evidently thought only a finger's worth of water in a puddle on the road wasn't very convincing; understandably so, given the condition of English roads at the time:

> Now in the highwayes litle shallow drills
> Which every raine amongst the pebbles fills,
> Scarce a foote deepe make th'heavens seeme as farre
> Beneath the ground, as they above it are . . .
> (*Lucy Hutchinson's Translation of Lucretius' De Rerum Natura*, ed. Hugh de Quehen, 1996)

11. *carpenter's rule*: In Epicurus' lost work the *Canon* (Greek for carpenter's rule or straight edge), he taught the reliability of sense-perception in determining false from true propositions.

12. *Those of the melancholy plaints . . . nightingales*: There is a serious textual corruption in the Helicon line, so I have permitted myself considerable licence. The melodious bird mentioned might instead be the swan (whose song Lucretius praises several times as being particularly pleasing, e.g. in lines 181 and 909a), but I chose nightingales, suggested by some editors, with their poetic connotations in English verse.

13. *A portion . . . upon the breeze*: Cf. Jesus's Parable of the Sower and the Seed (Matthew 13:3–9; Mark 4:3–9; Luke 8:5–8).

14. *Things are made of many seeds . . . combined*: Cf. I.814–29, 895, 896; II.333–80.

15. *bitter . . . now repeat*: Lucretius has mentioned the sweetness of honey before (I.939, II.397, IV.13), but not its bitterness. Democritus thought that honey was neither sweet nor bitter, Heraclitus that it was both. Of course, honey can certainly contain other flavours depending on what flowers are used by the bees.

16. *snow-white goose . . . Roman Citadel*: The honking of geese sacred to Juno on the Capitol in Rome saved it from the sack of the Gauls in 387 BC (see Livy 5.47.1–4).

17. *seems to shift its pose*: As the cells of an animated cartoon, or frames of film.

18. *For nothing is born . . . for them*: This approaches the concept in modern evolutionary theory of 'non-adaptive side consequences'; i.e., Lucretius asserts that the tongue did not evolve (or was not created) for talking; rather, speech is partly a side-effect of the

tongue. Lucretius is here arguing against Aristotle and the Stoics who held the teleological view. The debate about 'intelligent design' is an ancient one.

19. Lines 1000–1003 are a repetition of 992–5, and are omitted, 1000 actually being 1004.

20. *For those in adolescence's riptide . . . stains the sheet*: Hutchinson omits a hundred lines here, 1036–1136 (and glosses over other graphic passages), remarking on their contents: 'The cause and effects of Love which he makes a kind of dreame but much here was left out for a midwife to translate whose obsceane art it would better become than a nicer pen' (ed. de Quehen).

21. *Sicyon . . . Alindan silks . . . Cos*: The place-names associated with specific luxuries in the ancient world would not be unlike luxury 'branding' in our own time. Sicyon (in the Peloponnese in Greece) was famous for its fine (and effeminate) slippers. Lucretius' *Alidensia* probably refers to Alinda, a town in the province of Caria, in modern Turkey. He has *Cia* or *Cea*, but probably means Cos, a Greek island famous for a diaphanous fabric.

22. *She does – we know it's true . . . laughter*: Cf. Jonathan Swift's 'The Lady's Dressing Room', where the lover, upon discovering the chamber-pot in his mistress's boudoir, exclaims, 'Oh! Celia, Celia, Celia shits!'

23. *But the lover . . . the fool*: Lucretius is mocking the lover in the very language and props of love poetry itself, particularly the 'paraclausithyron', a standard situation in comedy and love poetry (and, no doubt, life) in which the lover, after a night of revelry (hence the 'wreaths in bloom', which were worn on the heads of drinkers), goes to his beloved's door in a maudlin state and petitions for entrance, often with a song. The unyielding door itself was sometimes addressed in such poems, hence the 'haughty doorposts' (Frank O. Copley, *Exclusus Amator: A Study in Latin Love Poetry*, 1956).

24. *parents carry elemental seeds . . . family tree*: This seems surprisingly close to a concept of genes.

BOOK V

COSMOS AND CIVILIZATION

1. *him*: Epicurus.

2. *calm waters*: The Greek term for the Epicurean doctrine of 'untroubledness' – *ataraxia*, tranquillity – is a metaphor from sailing on calm waters.

3. *there are tribes*: Lucretius is perhaps thinking of the Germans, of whom Caesar said, 'They do not concern themselves with agriculture; instead the greater part of their diet consists of milk, cheese and meat' (*Gallic War*, VI.22, trans. AES).

4. *deeds of Hercules*: The Twelve Labours of Hercules were: 1. slaying the Nemean lion; 2. slaying the Lernean hydra; 3. capture of the Arcadian stag; 4. destruction of the Erymanthian boar; 5. cleansing the Augean stables; 6. destruction of the Stymphalian birds; 7. capture of the Cretan bull; 8. capture of the man-eating mares of the Thracian Diomedes; 9. seizure of the girdle of the queen of the Amazons; 10. capture of the oxen of Geryon; 11. fetching the golden apples of the Hesperides; 12. bringing Cerberus from the underworld.

   Lucretius' dismissive treatment of Hercules is pointed. Hercules was the patron 'saint' of the Stoics, a mortal who through his Labours on behalf of mankind achieved divinity. Lucretius holds that the mortal Epicurus did more true good for mankind and is thus more deserving of godhead.

5. *Pride . . . Sloth*: D. E. W. Wormell thinks it is no accident that these, along with Avarice and Ambition, in III.59, correspond so closely to the Seven Deadly Sins (a correspondence emphasized in the translation). He suggests that the Church may have absorbed the list from Epicurean teaching ('The Personal World of Lucretius', in *Lucretius*, ed. D. R. Dudley (1965)).

6. *gods exist . . . world's zones*: Epicureanism held that the gods lived in the spaces between the worlds (*intermundia*).

7. *I'll demonstrate . . . soon enough*: Lucretius doesn't, however. It may be that he intended to and did not, or that this line refers not to the nature of the gods, but to something else.

8. *A human baby's like . . . woe*: A famous and widely imitated passage; Wordsworth in 'To ——, Upon the Birth of her First-Born Child' (lines 1–12):

> Like a shipwrecked Sailor tost
> By rough waves on a perilous coast,
> Lies the Babe, in helplessness
> And in tenderest nakedness,
> Flung by labouring Nature forth
> Upon the mercies of the earth.
> Can its eyes beseech? no more
> Than the hands are free to implore:
> Voice but serves for one brief cry;
> Plaint was it? or prophesy
> Of sorrow that will surely come?
> Omen of man's grievous doom!

The poem has an epigraph: 'Tum porro puer, ut saevis projectus ab undis / Navita, nudus humi jacet, etc.' – LUCRETIUS.

9.   *They do not need a change of clothes . . . no dearth*: Hard not to be put in mind here of the Sermon on the Mount: 'And why take ye thought for raiment? Consider the lilies of the field, how they grow; they toil not, neither do they spin. And yet I say unto you, that even Solomon in all his glory was not arrayed like one of these' (Matthew 6:28–9 in King James version).

10. *earth . . . heat*: Earth, air, water, fire – the four 'roots' of matter according to Empedocles.

11. *major members*: The four 'elements' – earth, air, water, fire.

12. *salt . . . filtered out*: Cf. II.472–4.

13. *Look upon my works ye mighty, and despair!*: The line itself is corrupt, which gave the translator more than the usual licence. As its ironic tone and context strongly suggest the famous line from Shelley's 'Ozymandias', the translator found appropriating it irresistible.

14. *Events . . . Troy*: Referring to the *Iliad* and to a lost epic poem, the *Thebaïs*, which told the story of the *Seven Against Thebes* (later a play by Aeschylus) – the seven who fought unsuccessfully to restore Polynices (son of Oedipus and brother to Eteocles) to the throne of Thebes.

15. *these things*: I.e., sky, sun, earth, sea.

16. *major members of the world*: Empedocles' four 'roots' of matter – earth, air, water, fire – which here form four members of one 'state' – the world – warring amongst themselves.

17. *legends tell*: The story is of Deucalion. For the Greek version of the Great Flood, and of Deucalion (the Greek Noah) and his wife, Pyrrha, see Ovid, *Metamorphoses* 1.253–415.

18. *Strife*: Lucretius seems to have in mind the Empedoclean notion of Strife here; see Introduction, p. xii.

19. *much of the fire*: Not all, since some remained in the earth, as seen in volcanoes, etc.

20. *Black Sea . . . its tide*: A belief widely held in ancient times, that the Pontus – the Black Sea – flowed always into the Propontis – the Sea of Marmara – towards the Aegean. Cf. Shakespeare's *Othello* (III.3.450–53):

> Like to the Pontick sea,
> Whose icy current and compulsive course
> Ne'er feels retiring ebb, but keeps due on
> To the Propontic and the Hellespont.

21. *outside air . . . pole*: The air holds the axis in place so that the globe can spin (as with a model globe of the earth).

22. *dawn's shining standard . . . coming of the day*: Cf. A. E. Housman's poem, 'Revolution': 'Day's beamy banner up the east is borne'.

23. *teaching of the Babylonians*: Specifically, the teachings of Berosus (*fl. c.* 260 BC). The Babylonians were famed astrologers.

24. *Spring . . . brings up the rear*: This description of the procession of Spring is a likely influence on Botticelli's famous *Allegoria della Primavera* (*c.* 1482). Cupid is the winged god of love, son of Venus. Flora was the Roman goddess of flowers and spring. Zephyr is the warm West Wind. The summer Northerlies are the Meltemia, hot winds that kick up in the Mediterranean in the dog-days of August. Volturnus is the east-southeast wind. For Winter, cf. Spenser, *The Faerie Queene* 7.31.1–2: 'Lastly, came Winter cloathed all in frize, / Chattering his teeth for cold that did him chill'.

25. *shady cone*: The shadow cast by the earth. Cf. Housman's 'Revolution':

> But over sea and continent from sight
> Safe to the Indies has the earth conveyed
> The vast and moon-eclipsing cone of night,
> Her towering foolscap of eternal shade.

26. *hemlock . . . sleek and wide*: This statement turns out to be untrue – goats are indeed susceptible to the poisonous plant hemlock. Hemlock is famously associated with the death of Socrates; he carried out his own execution by drinking a draught of hemlock in prison.

27. *boar . . . trees*: An observation from nature: boars do indeed make nests for themselves, tunnels of leaves and branches.

28. *perverse science of navigation . . . gloom*: Romans were suspicious of sailing and navigation, as comes across again and again in this book.

29. *Molossian mastiff*: Greece was famous for two breeds of dogs: the Laconian, a slighter, swift hunting dog, from Sparta, and the Molossian, a heavier shepherd/guard dog from Epirus in the north of Greece, which gave rise to the modern mastiffs. There is a larger-than-life-size statue of the Molossian in the British Museum. A native breed directly descended from the Molossian, the Greek Shepherd, can be seen working in Epirus to this day. They are all white, though through interbreeding they are somewhat smaller than their ancestors, and are still renowned for a fierce disposition.

30. *men got fire from either one of these*: Lightning or friction.

31. *take the commander . . . elephants of war*: Pyrrhus, sailing to Tarentum (in the heel of Italy) with a fleet carrying twenty elephants (as well as 3,000 horse, 20,000 foot, etc.), was caught in an unseasonable squall. Most of the fleet was lost, scattered and driven upon rocks, but Pyrrhus himself made it to land by diving overboard and swimming to shore (see Plutarch, *Life of Pyrrhus*). See also Book II, note 13.

32. *glorious rods . . . of power*: Referring to the fasces, a bundle of rods of elm or birch bound with red thongs and attached to an axe. They symbolized authority and were carried by attendants known as lictors before Roman officials.

33. *Some also experimented . . . laying many low*: On this section, lines 1308–40, Bailey in his Commentary wonders, 'Whence did Lucr. obtain these strange stories?' He goes on to say, 'This paragraph, more than anything else in the poem, makes me wonder whether Jerome was not right, and that Lucr.'s mind was from time to time deranged.' St Jerome, writing at the end of the fourth century AD, asserts, in an account that is now given little credence, that Lucretius wrote his books during spells of insanity; in fact, such scenes would not have been a great leap of imagination for a Roman of the first century BC. *Venationes*, fights between men and beasts, or beasts and beasts, including wild boar, lions, leopards, elephants, etc. (and sometimes including cavalry), were staged as opening acts for gladiatorial games (see Book VI, note 8).
    struck with a botched blow . . . laying many low: Carthaginian mahouts were equipped with an iron spike to drive into the brain of the elephant if it was wounded, for a wounded elephant would

run amok. There is a historical example of an elephant killing friends as well as foes found in Plutarch's *Life of Pyrrhus*:

> Another elephant named Nicon, one of those which had advanced further into the city, was trying to find its rider who had been wounded and fallen off its back, and was battling against the tide of fugitives who were trying to escape. The beast crushed friend and foe together indiscriminately until, having found its master's dead body, it lifted the corpse with its trunk, laid it across its tusks, and wheeling round in a frenzy of grief, turned back, trampling and killing all who stood in his path. (*Plutarch: The Age of Alexander*, trans. Ian Scott-Kilvert, Penguin Classics, 1973)

34. *smooth enough*: Rough edges on a loom would snag or cut the yarn. Loom-weaving clearly pre-dates the Iron Age, however.
35. *[Thus Time . . . shore]*: Possibly an interpolation into the text. These lines are repeated at the end of this book at 1454–5.

BOOK VI

WEATHER AND THE EARTH

1. *ailing mortals*: 'Wretched' is the traditional Homeric epithet for mortals (*deiloisi brotoisi*), but Lucretius uses a Latin word (*aegris*) that means both wretched and sickly, thus having it do double duty, foreshadowing the plague that will descend upon the Athenians indiscriminately at the end of the book.
2. *Athens the Illustrious . . . men*: Praise of Athens as the giver of great things to mankind was conventional. Regarding agriculture, Lucretius may have had in mind the myth of Demeter giving the art of agriculture to Triptolemus (who was from Eleusis, near Athens); regarding laws and governance, Lucretius may have had in mind Draco (by tradition, the author of the first written code of law for Athens (621 BC) – most sentences were death, hence 'Draconian') and Solon (born 639 BC, who remodelled the Athenian constitution, repealing Draco's harsh measures except those regarding homicide).
3. *a genius of a man*: Epicurus.
4. *[Because . . . divine]*: As at I.153–4. It may be an interpolation here, and in 90–91, where the lines are repeated.
5. *Section . . . in different bits*: Etruscan augury divided the sky into sixteen sections (according to Cicero, *De Divinatione* II.xviii.42).
6. *like a tent*: See Book IV, note 2 for awnings over Roman theatres.

7. *Or when a buffeting breeze . . . flapping by*: It is tempting to see this as a brief glimpse of the poet himself at work (as at I.142, where he is writing late into the night), maybe in the garden, with his papers blown from his table!

8. *winds . . . cage*: An observation from life. Lucretius would doubtless have seen wild animals – as lions and leopards – caged for the games. *Venationes* (spectacles involving men against beasts or beasts against beasts) were introduced into Rome in the second century BC and quickly became more extravagant and elaborate. Pompey staged an event in 55 BC in which were slaughtered some 600 lions, 410 leopards, as well a score of elephants (whose plight and evident intelligence excited the pity of the crowd); see George Jennison, *Animals for Show and Pleasure in Ancient Rome* (1937).

9. *the Great Flood*: Cf. Book V, note 18.

10. *Etruscan scrolls*: The Etruscan *libri fulgurales* (books of the thunder-bolts) discussed the interpretation of thunder and lightning. Other Etruscan books governed rituals and divination by entrails. It was a highly codified religion, and superstitious even by ancient standards.

11. *bolt . . . pollute a place*: Referring to a 'bidental' – a place struck by lightning, or where a person has been killed by lightning. People killed by lightning had to be buried where they fell. Places were polluted and had to be purified by sacrifice of a *bidens* (Latin for a two-year-old sheep), a superstition of Etruscan origin.

12. *other deities*: Etruscan religion held that various gods, not just Jove/ Jupiter (king of the gods, equivalent of Greek Zeus), were able to hurl thunderbolts. 'The Etruscan Jupiter . . . threw three kinds of thunderbolts either mild or more or less devastating; eight other gods threw one kind each' (*Oxford Classical Dictionary*, 'Religion, Etruscan').

13. *Sidon . . . Aegium*: The earthquake at Sidon probably occurred late in the fifth century BC. The towns of Helice and Buris, near Aegium (on the southwest shore of the Gulf of Corinth), were swallowed up by the sea during an earthquake in 373 BC. Ancient Helice, now under water, was only recently rediscovered by archaeologists who are in the process of excavating it.

14. *Men wonder*: The abrupt shift here may indicate lack of revision or a lacuna.

15. *Sacred Fire*: The disease erysipelas or St Anthony's Fire.

16. *krater*: The Greek name for a large vessel for mixing wine with water when serving. Likewise, Lucretius suggests, the crater of the volcano is a 'Mixing bowl' for sand, water, rocks, etc.

17. *yearly winds*: Northerly winds that (even today) kick up after the

middle of August in the eastern Aegean (the *Meltemia* in modern Greek).

18. *Etesian*: A dry northwesterly summer wind in the eastern Mediter-
     ranean.
19. *Avernian*: Folk etymology of 'Avernus' was from the Greek *aornos*,
     meaning 'bird-less'. Lucretius calls all bird-less places Avernian,
     after Lake Avernus (now Averno), a lake in Italy near Naples, which
     gives off sulphuric fumes and was supposed to be an entrance to the
     underworld.
20. *Tritonian*: Cult name of Athena, who in one account was born on
     Lake Tritonis in Libya. Herodotus (*History* 4.189) adds that her
     attributes, the aegis (goat buckler) and the robe for her cult statue,
     were Libyan in origin, resembling the traditional gear of Libyan
     women.
21. *croaking crows . . . spying*: The daughters of Cecrops (first king of
     Athens) disobeyed Athena's command not to open the chest that
     contained the baby Erichthonius. It was the crow that tattled on
     them and Athena punished crows for their excessive vigilance by
     banning them from the Acropolis.
22. *There is another spot . . . quadrupeds*: Strabo mentions in his *Geog-
     raphy* (XIII.630) a plutonium (i.e., avernian, a spot with mephitic
     vapours) at Hierapolis (modern Manbij), near Laodicea, and says:

> Now to those who approach the handrail anywhere round the enclos-
> ure the air is harmless, since the outside is free from that vapour in
> calm weather, for the vapour then stays inside the enclosure, but any
> animal that passes inside meets instant death. At any rate, bulls that
> are led into it fall and are dragged out dead; and I threw in sparrows
> and they immediately breathed their last and fell. (trans. Horace Leon-
> ard Jones, 1929)

23. *sacrificed to spirits down below*: See III.52.
24. *wing-footed stags . . . holes*: Regarding this curious belief, Pliny the
     Elder in his *Natural History* relates: 'No one is ignorant of the fact
     that deer are the bane of serpents, and, if they happen upon any,
     draw them out of their holes and eat them' (28.42.149).
25. *certain trees . . . beneath the canopy*: The idea that lying under a
     tree can give one a headache will come as no surprise to allergy
     sufferers! Pliny in his *Natural History* mentions the walnut, a tree
     with oppressive shade (17.89), and Virgil in his *Eclogues* (10.76)
     mentions the juniper.
26. *Helicon . . . man's demise*: The tree on Helicon whose shade is

supposedly fatal has not been identified, but Pliny attributes this power to the yew tree in Arcadia (*Natural History* 16.51).

27. *falling fits . . . mouth*: Epilepsy.

28. *A woman . . . faint*: Castor musk is a pungent extraction from beaver glands used for medicinal purposes. The idea that menstruating women will faint at certain odours may seem odd, but certainly many pregnant women are sensitized to strong smells.

29. *hot sun . . . pall of black*: The idea being that the sun goes under the ground during the night. Indeed, Herodotus mentions that the spring is known as 'the spring of the sun' (*History* 4.181).

30. *just-extinguished wick . . . flame*: This is an easy experiment to perform with a match and a candle. A still-smoking wick will indeed light at a distance from the flame of the match if the match touches the wisp of smoke.

31. *Magnesia*: Magnesia ad Sipylum, a city of Lydia (inland western Asia Minor), founded by the Aeolian Magnetes of eastern Thessaly. A 'magnet' is literally a stone from Magnesia. See also next note.

32. *rings . . . course*: Socrates, in Plato's *Ion*, uses the chain of rings suspended from a magnet as a metaphor for the divine force of inspiration. '. . . this is not an art in you, whereby you speak well on Homer, but a divine power, which moves you like that in the stone which Euripides named a magnet, but most people call "Heraclea stone." For this stone not only attracts iron rings, but also imparts to them a power whereby they in turn are able to do the very same thing as the stone, and attract other rings; so that sometimes there is formed quite a long chain of bits of iron and rings, suspended one from another; and they all depend for this power on that one stone' (trans. by W. R. M. Lamb, 1925). Heraclea was to the south of Magnesia. *Magnetis lithos*, 'magnetic stone', appears in a fragment of Euripides' *Oeneus*.

33. *Samothracian irons*: Possibly amulets (for which the Aegean island Samothrace was known) consisting of iron rings.

34. *Wine-dark stain of mollusks*: See II.501 and note.

35. *chrysocolla*: 'Gold glue', often identified with borax.

36. *tide of pestilence . . . Athens*: The account of the plague at Athens which closes Book VI and the poem comes from the reporting of the plague by Thucydides (430 BC), in his *History of the Peloponnesian War*, Book 2. (Thucydides was one of the survivors of the plague.) While remaining very close to Thucydides' account, Lucretius does alter some telling details.

37. *clan . . . hardest hit*: Concern for the plight of the dogs is a typically Lucretian touch. Thucydides in his *History of the Peloponnesian*

War only mentions dogs to say that 'dogs, being domestic animals, provided the best opportunity of observing this effect of the plague' (trans. Rex Warner, 2004).

38. *desert their dead*: In almost an embodiment of Paul Valéry's famous assertion that 'a poem is never finished, only abandoned', the poem ends with desertion; in Latin the verb *desererentur* is the last word of the poem as it stands. There has been much discussion about the poem's closure, or lack thereof. Many think this bleak ending (with its focus on the corpses of the plague) to a poem against the fear of death cannot be what Lucretius, who died while the poem was in an unrevised state, intended (see Introduction, p. viii); but it has an undeniable poetic power for the modern reader.

Glossary of Proper Names

(All dates are BC)

Achaea a district of Greece in the south of Thessaly.

Acheron originally a river of the underworld, becoming a name for the underworld itself.

Aetna active volcanic mountain in northeast Sicily. Notable eruptions in ancient times occurred in 475 (alluded to in Aeschylus and Pindar), 425 (mentioned by Thucydides), 396 and 122.

Ammon Egyptian divinity identified with Zeus/Jupiter. His temple and spring were in the Ammonium oasis in the Libyan desert. Herodotus describes the spring's properties (cold at dawn, warm at midday, boiling at midnight, then cool at dawn again) in his *History* 4.181.

Anaxagoras philosopher born in 500 at Clazomenae in Ionia, who taught at Athens. His students included Pericles and Euripides. For his philosophy of homoeomery, see I.830 ff. and note.

Apollo god of, among other things, medicine, music (he is depicted with a lyre), archery and prophecy. The centre of his worship was at Delphi, where the Sibyl or Pythia, his priestess, would deliver oracles while seated on a bronze tripod. He is also closely connected with the Muses. The laurel (or bay) is sacred to him.

Aradus island off the coast of Phoenicia – modern Ruad or Arwad – two miles off the coast of Tortosa, Syria.

Athena (Roman form, Minerva, but the Greek is retained because of her association with Athens) daughter of Zeus, warrior virgin-goddess and patron of the city of Athens. The Parthenon is her temple. One of her epithets is 'Glaucopis' – grey-green eyed (as in IV.1161).

Attica division of Greece containing Athens and its environs. A triangular promontory in the eastern central region, it is bounded by the Aegean sea, the Saronic gulf and Boeotia.

Bacchus see Dionysus.

Calliope muse of epic poetry. Her name means 'she of the lovely voice'.

Carthage city-state founded by the Phoenicians on the northern coast of Africa. Carthage was the great rival of Rome, and they fought three wars. The Second Punic War (218–201) involved Hannibal's daring invasion of Italy by crossing the Alps with a contingent of war elephants. Carthage was destroyed in the Third Punic War in 146.

Cecrops traditionally the first king of Attica and founder of Athens.

Centaurs mythical race of half-men, half-horse creatures, the offspring of Ixion and a cloud, who inhabited Mount Pelion in Thessaly.

Cerberus three-headed hound that guarded the underworld.

Ceres (Greek, Demeter) goddess of grain and the earth. She gave agriculture to mankind through Triptolemus, son of Celeus, king of Eleusis (a town to the northwest of Athens on the coast). He received seeds of wheat and a chariot drawn by winged dragons, and flew over the earth in his chariot bringing agriculture to mankind.

Charybdis see Scylla.

Chimaera fire-breathing monster slain by Bellerophon, it was made up of three creatures: from head to middle to tail, a lion, a goat and a serpent respectively.

Cumae town founded as the earliest Greek colony in Italy (*c.* 750), on the coast near modern-day Naples. It was here that the Sibyl, a prophetess, had her cavern, as in Virgil's *Aeneid*, Book 6, and it was associated with Lake Avernus and the entrance to the underworld.

Cupid (Greek, Eros) winged god of love, son of Venus.

Cybele the Great Mother, a fertility mother-goddess of Anatolia, associated with the Mother Earth and with Cretan goddess Rhea (mother of Jupiter/Zeus). Her worship in Rome, introduced from Phrygia in 205–204, was restricted in Lucretius' time – participation in the priesthood and rituals (except for the procession) were forbidden to Roman citizens. She was associated with Mount Ida in Asia Minor (another reason for her conflation with Rhea).

Democritus Greek philosopher born *c.* 460 at Abdera in Thrace, who developed the atomic theory of Leucippus, his teacher. (He is an important precursor to Epicurus.) He was known as 'the laughing philosopher' for his cheerful outlook. Legend has it that he blinded himself to avoid distractions. He lived to over 100 (perhaps even 109). Lucretius' is the oldest account we have that he died by suicide (III.1039–41). Later sources say he achieved this by refusing sustenance.

Dionysus/Bacchus god of wine.

Empedocles philosopher, poet and mystic of Acragas in Sicily (*c.* 493–*c.* 433). Two long hexameter poems are ascribed to him, *On Nature* and *Purifications*. His excellent poetry is the verse model for Lucretius, who sometimes quotes or echoes him (e.g. at II.7, 8; II.387, 388; III.343; V.102, 103; V.837 ff.), though, as we only have fragments of Empedocles, no doubt there are many echoes lost on us. Empedocles taught that there were four roots or elements – air, water, earth, fire – from which all things were formed: they were mixed and separated under the opposing forces of Love and Strife (see Introduction). Legend has it that he threw himself into the flames of Mount Aetna in a bid to be believed a god.

Ennius regarded as the father of Roman poetry. We only have fragments of his work. His most important poem was an epic history called the *Annales*, which opened with a dream in which Homer appears to Ennius and tells him he is Homer's own reincarnation. (The reincarnated soul supposedly passed first through a peacock, then Euphorbus, Homer, Pythagoras and Ennius.)

Epicurus Greek philosopher born in 342 on the island of Samos, later a resident of Athens (306), where he taught philosophy in his famed garden. He founded the Epicurean school of philosophy, that expounded by Lucretius' poem. He died in 270 at the age of seventy-two.

Furies, the three goddesses of vengeance, who lived in the depths of Tartarus.

Giants the Gigantes were born of the Earth. They revolted against Heaven (their 'crime', V.118–19) but were defeated by Hercules and the gods and were buried under Aetna and other volcanoes.

Helen daughter of Leda and Zeus, renowned for her beauty; she was the wife of Menelaus and queen of Sparta. Her abduction by the Trojan prince Paris (aka Alexander) to Troy, sets the Trojan War into motion.

Helicon mountain of Boeotia (district of central Greece), sacred to Apollo and the Muses. It is home of the sacred springs Aganippe and Hippocrene, said to be sources of inspiration.

Heraclitus Presocratic philosopher (*c.* 540–*c.* 480) who wrote a work entitled *On Nature* and famous for such aphorisms as 'everything flows'. He taught that fire was the primary form of all matter.

Ida mountain in central Crete where Zeus was said to have been hidden as an infant in a cave. Also, a mountain range in Phrygia (Asia Minor) connected with the worship of Cybele. V.663 refers to Mount Ida in Phrygia.

Jove see Jupiter.

Juno wife of Jupiter, queen of the gods, identified with Greek Hera.

Jupiter/Jove two forms of the name of the Roman sky-god, identified with Greek Zeus, king of the gods.

Mars Roman god of war and strife, identified with Greek Ares. Originally an agricultural deity, a god of fields, he becomes god of war since fields are where wars are fought. He is Venus' lover and also the father of Remus and Romulus, the latter being the legendary founder and first king of Rome. In a sense, Venus and Mars are the 'parents' of Rome.

Matuta ancient Roman goddess of first morning light (as well as child-birth), identified with Greek Leucothea and Roman Aurora.

Memmius Gaius Memmius, the dedicatee of the poem, was a poetic patron, minor poet (none of his work survives) and orator. He was governor of Bithynia (district of Asia Minor, in modern Turkey) in 58 and is mentioned in unflattering terms by Lucretius' contemporary, Catullus. Banished in 53 in the wake of an electoral scandal, he went into exile in Athens. It is unclear what his relationship is to the Epicurean school. He owned the estate on which were the ruins of Epicurus' house and Epicurus' disciples strongly objected to his plans to build on the site; Cicero persuaded him to give up the project.

Muses goddesses (traditionally nine in number) of inspiration and the arts, the daughters of Zeus and Mnemosyne. They were associated with mountains and springs, particularly Mount Helicon in Boeotia, with its famous spring the Hippocrene. They are associated with bees, who also favour mountainsides, and their sweet productions of honey. See also Calliope.

Neptune Roman god of the sea (Greek Poseidon); sometimes the sea itself.

Orcus ancient Roman god of the underworld; sometimes the underworld itself.

Pan Greek god of flocks and shepherds, represented with horns and goat's feet, inventor of the shepherd's flute (pan pipe). The Romans identified him with Faunus, an ancient Italian deity, guardian of flocks, but also responsible for spooky sights and mysterious sounds in the forest. He was said to instil irrational fear in herds or men, called after him a 'panic'. Faunus was sometimes thought of as many demigods – i.e., fauns.

Pandion name of two early kings of Attica (either the son of Erichthonius, father of Procne and Philomela, or the son of Cecrops, later expelled from Athens).

Phaëthon mortal son of the sun-god, Phoebus (Helios), who persuaded his father to let him drive the team of horses of the sun (an early version of a teenager asking for the keys to the car), but didn't have the strength to control the reins. The horses ran wildly out of control, the sun coming so close to the earth that it threatened to burn it up. Zeus struck him from the chariot with a thunderbolt. (See Ovid, *Metamorphoses* 1.750–2.400.)

Pythia see Apollo.

Rhea ancient Greek goddess of the earth. She is the wife of Cronus (Roman Saturn) and mother of Hestia, Demeter, Hera, Hades, Posei-don and Zeus (Vesta, Ceres, Juno, Pluto, Neptune and Jupiter accordingly in their Roman equivalents). The seat of her worship was in Crete. She is sometimes conflated with Cybele, another mother-goddess.

Saturn (identified with Greek Cronus) son of Heaven and Earth, husband of Rhea, and father of Jupiter.

Scylla and Charybdis two treacherous rocks between Sicily and Italy. In ancient mythology, Scylla was a monster who lived in a cave near the rock on the Italian side; she had twelve feet, and six necks and heads, each of which contained three rows of sharp teeth. Below the waist she was made of barking dog-like creatures. Charybdis swallowed water and vomited it up again, causing a violent whirlpool.

Sibyl see Apollo.

Sisyphus king who founded Corinth. He was associated with trickery (accounts of his crimes vary). His punishment in the underworld was to roll uphill a marble boulder that always rolled back once he reached the top.

Skapte Hyle town in Thrace (northern Greece) noted for its gold mines. (Thucydides, who had property in the mines, was said to have composed some of his history in exile there.)

Tartarus region of the underworld reserved for punishments – thus, the underworld itself, Hell.

Tityus a giant (son of Gaea, the Earth), punished in the underworld for attempted rape (of Leto in Homer, but accounts vary). His punishment in Tartarus was to lie stretched on the ground over nine acres while two vultures perpetually ate his liver.

Venus goddess of Love, identified with Greek Aphrodite, she was the mother of the Trojan hero Aeneas, legendary ancestor of the Romans. Her association with sexual love makes her a symbol of Nature's fertility and abundance. (Venus' image was also stamped on coins minted by the gens Memmia, the clan to which Memmius belonged.) Her lover is traditionally Mars, god of war. Lucretius' use of their pairing also suggests the Empedoclean notion of the rival forces of Love and Strife (see Introduction). 'Venus' also sometimes refers to the act of love or love itself.

PENGUIN CLASSICS

THE ODYSSEY   HOMER

'I long to reach my home and see the day of my return. It is my never-failing wish'

The epic tale of Odysseus and his ten-year journey home after the Trojan War forms one of the earliest and greatest works of Western literature. Confronted by natural and supernatural threats – shipwrecks, battles, monsters and the implacable enmity of the sea-god Poseidon – Odysseus must test his bravery and native cunning to the full if he is to reach his homeland safely and overcome the obstacles that, even there, await him.

E. V. Rieu's translation of *The Odyssey* was the very first Penguin Classic to be published, and has itself achieved classic status. For this edition, his text has been sensitively revised and a new introduction added to complement E. V. Rieu's original introduction.

'One of the world's most vital tales ... *The Odyssey* remains central to literature' Malcolm Bradbury

Translated by E. V. Rieu
Revised translation by D. C. H. Rieu, with an introduction by Peter Jones

PENGUIN CLASSICS

THE ILIAD   HOMER

'Look at me. I am the son of a great man. A goddess was my mother. Yet death and inexorable destiny are waiting for me'

One of the foremost achievements in Western literature, Homer's *Iliad* tells the story of the darkest episode in the Trojan War. At its centre is Achilles, the greatest warrior-champion of the Greeks, and his refusal to fight after being humiliated by his leader Agamemnon. But when the Trojan Hector kills Achilles's close friend Patroclus, he storms back into battle to take revenge – although knowing this will ensure his own early death. Interwoven with this tragic sequence of events are powerfully moving descriptions of the ebb and flow of battle, of the domestic world inside Troy's besieged city of Ilium, and of the conflicts between the gods on Olympus as they argue over the fate of mortals.

E. V. Rieu's acclaimed translation of Homer's *Iliad* was one of the first titles published in Penguin Classics, and now has classic status itself. For this edition, Rieu's text has been revised, and a new introduction and notes by Peter Jones complement the original introduction.

Translated by E. V. Rieu
Revised and updated by Peter Jones with D. C. H. Rieu
Edited with an introduction and notes by Peter Jones

PENGUIN CLASSICS

THE RISE OF THE ROMAN EMPIRE   POLYBIUS

'If history is deprived of the truth, we are left with nothing but an idle, unprofitable tale.'

In writing his account of the relentless growth of the Roman Empire, the Greek statesman Polybius (*c*. 200–118 BC) set out to help his fellow-countrymen understand how their world came to be dominated by Rome. Opening with the Punic War in 264 BC, he vividly records the critical stages of Roman expansion: its campaigns throughout the Mediterranean, the temporary setbacks inflicted by Hannibal and the final destruction of Carthage in 146 BC. An active participant in contemporary politics, as well as a friend of many prominent Roman citizens, Polybius was able to draw on a range of eyewitness accounts and on his own experiences of many of the central events, giving his work immediacy and authority.

Ian Scott-Kilvert's translation fully preserves the clarity of Polybius's narrative. This substantial selection of the surviving volumes is accompanied by an introduction by F. W. Walbank, which examines Polybius's life and times, and the sources and technique he employed in writing his history.

Translated by Ian Scott-Kilvert
Selected with an introduction by F. W. Walbank

PENGUIN CLASSICS

THE CAMPAIGNS OF ALEXANDER   ARRIAN

'His passion was for glory only, and in that he was insatiable'

Although written over four hundred years after Alexander's death, Arrian's *Campaigns of Alexander* is the most reliable account of the man and his achievements we have. Arrian's own experience as a military commander gave him unique insights into the life of the world's greatest conqueror. He tells of Alexander's violent suppression of the Theban rebellion, his total defeat of Persia, and his campaigns through Egypt, India and Babylon – establishing new cities and destroying others in his path. While Alexander emerges from this record as an unparalleled and charismatic leader, Arrian succeeds brilliantly in creating an objective and fully rounded portrait of a man of boundless ambition, who was exposed to the temptations of power and worshipped as a god in his own lifetime.

Aubrey de Sélincourt's vivid translation is accompanied by J. R. Hamilton's introduction, which discusses Arrian's life and times, his synthesis of other classical sources and the composition of Alexander's army. The edition also includes maps, a list for further reading and a detailed index.

Translated by Aubrey de Sélincourt
Revised, with a new introduction and notes by J. R. Hamilton

PENGUIN CLASSICS

THE ANNALS OF IMPERIAL ROME   TACITUS

'Nero was already corrupted by every lust, natural and unnatural'

The Annals of Imperial Rome recount the major historical events from the years shortly before the death of Augustus to the death of Nero in AD 68. With clarity and vivid intensity Tacitus describes the reign of terror under the corrupt Tiberius, the great fire of Rome during the time of Nero and the wars, poisonings, scandals, conspiracies and murders that were part of imperial life. Despite his claim that the *Annals* were written objectively, Tacitus's account is sharply critical of the emperors' excesses and fearful for the future of imperial Rome, while also filled with a longing for its past glories.

Michael Grant's fine translation captures the moral tone, astringent wit and stylish vigour of the original. His introduction discusses the life and works of Tacitus and the historical context of the *Annals*. This edition also contains a key to place names and technical terms, maps, tables and suggestions for further reading.

Translated with an introduction by Michael Grant

PENGUIN CLASSICS

METAMORPHOSES OVID

'Her soft white bosom was ringed in a layer
of bark, her hair was turned into foliage, her arms into branches'

Ovid's sensuous and witty poem brings together a dazzling array of
mythological tales, ingeniously linked by the idea of transformation –
often as a result of love or lust – where men and women find themselves
magically changed into new and sometimes extraordinary beings.
Beginning with the creation of the world and ending with the deification
of Augustus, Ovid interweaves many of the best-known myths and
legends of ancient Greece and Rome, including the stories of Daedalus
and Icarus, Pyramus and Thisbe, Pygmalion, Perseus and Andromeda,
and the Fall of Troy. Erudite but light-hearted, dramatic and yet playful,
the *Metamorphoses* has influenced writers and artists throughout the
centuries from Shakespeare and Titian to Picasso and Ted Hughes.

This lively, accessible new translation by David Raeburn is in hexameter
verse form, which brilliantly captures the energy and spontaneity of the
original. The edition contains an introduction discussing the life and work
of Ovid as well as a preface to each book, explanatory notes and an index
of people, gods and places.

A new verse translation by David Raeburn with an introduction by
Denis Feeney

THE STORY OF PENGUIN CLASSICS

Before 1946 ... 'Classics' are mainly the domain of academics and students; readable editions for everyone else are almost unheard of. This all changes when a little-known classicist, E. V. Rieu, presents Penguin founder Allen Lane with the translation of Homer's *Odyssey* that he has been working on in his spare time.

1946 Penguin Classics debuts with *The Odyssey*, which promptly sells three million copies. Suddenly, classics are no longer for the privileged few.

1950s Rieu, now series editor, turns to professional writers for the best modern, readable translations, including Dorothy L. Sayers's *Inferno* and Robert Graves's unexpurgated *Twelve Caesars*.

1960s The Classics are given the distinctive black covers that have remained a constant throughout the life of the series. Rieu retires in 1964, hailing the Penguin Classics list as 'the greatest educative force of the twentieth century.'

1970s A new generation of translators swells the Penguin Classics ranks, introducing readers of English to classics of world literature from more than twenty languages. The list grows to encompass more history, philosophy, science, religion and politics.

1980s The Penguin American Library launches with titles such as *Uncle Tom's Cabin*, and joins forces with Penguin Classics to provide the most comprehensive library of world literature available from any paperback publisher.

1990s The launch of Penguin Audiobooks brings the classics to a listening audience for the first time, and in 1999 the worldwide launch of the Penguin Classics website extends their reach to the global online community.

The 21st Century Penguin Classics are completely redesigned for the first time in nearly twenty years. This world-famous series now consists of more than 1300 titles, making the widest range of the best books ever written available to millions – and constantly redefining what makes a 'classic'.

The Odyssey continues ...

The best books ever written

PENGUIN CLASSICS

SINCE 1946

Find out more at www.penguinclassics.com